Eagan Press Handbook Series

Colorants

F. J. Francis

eagan press
St. Paul, Minnesota, USA

Cover: Liquid dye (center picture) and annatto (flower) courtesy of Warner-Jenkinson Co., Inc.; pie and cupcake ©1997 Artville, LLC; candy courtesy of McCormick Flavors, Inc.; LabScan XE courtesy of HunterLab.

Library of Congress Catalog Card Number: 98-89677
International Standard Book Number: 1-891127-00-4

©1999 by the American Association of Cereal Chemists, Inc.

All rights reserved.
No part of this book may be reproduced in any form, including photocopy, microfilm, information storage and retrieval system, computer database or software, or by any means, including electronic or mechanical, without written permission from the publisher.

Reference in this publication to a trademark, proprietary product, or company name is intended for explicit description only and does not imply approval or recommendation of the product to the exclusion of others that may be suitable.

Printed in the United States of America on acid-free paper

American Association of Cereal Chemists
3340 Pilot Knob Road
St. Paul, Minnesota 55121-2097, USA

About the Eagan Press Handbook Series

The Eagan Press Handbook series was developed for food industry practitioners. It offers a practical approach to understanding the basics of food ingredients, applications, and processes—whether the reader is a research chemist wanting practical information compiled in a single source or a purchasing agent trying to understand product specifications. The handbook series is designed to reach a broad readership; the books are not limited to a single product category but rather serve professionals in all segments of the food processing industry and their allied suppliers.

In developing this series, Eagan Press recognized the need to fill the gap between the highly fragmented, theoretical, and often not readily available information in the scientific literature and the product-specific information available from suppliers. It enlisted experts in specific areas to contribute their expertise to the development and fruition of this series.

The content of the books has been prepared in a rigorous manner, including substantial peer review and editing, and is presented in a user friendly format with definitions of terms, examples, illustrations, and trouble-shooting tips. The result is a set of practical guides containing information useful to those involved in product development, production, testing, ingredient purchasing, engineering, and marketing aspects of the food industry.

Acknowledgment of Sponsors for *Colorants*

Eagan Press would like to thank the following companies for their financial support of this handbook:

> Monarch Food Colors
> High Ridge, Missouri
> 314/677-6622
>
> Warner-Jenkinson Company, Inc.
> St. Louis, Missouri
> 800/325-8110

Eagan Press has designed this handbook series as practical guides serving the interests of the food industry as a whole rather than the individual interests of any single company. Nonetheless, corporate sponsorship has allowed these books to be more affordable for a wide audience.

Acknowledgments

Eagan Press thanks the following individuals for their contributions to the preparation of this book:

Penny Huck and other staff at Warner Jenkinson Co. Inc., St. Louis, MO, for contributing chapters 12–15 and providing information on regulations.

Winston A. Boyd, Color Craft Consulting, Schaumburg, IL, for contributing the troubleshooting guides and reviewing the manuscript.

Eagan Press also appreciates the help of the following reviewers:

Per P. Isagor, Chr. Hansen, Inc., Milwaukee, WI

Jerry Kinnison, Warner-Jenkinson Co., St. Louis, MO

Gabe Lauro, Natural Color Resource Center, Orange, CA

Gayle Stout, Quest International, Hoffman Estates, IL

Contents

1. **Overview • 1**
   Adulteration of Foods
   Appreciation of Color

2. **Measurement of Color • 7**
   Physiological Basis of Color
   Visual Systems
   Spectrophotometric Measurement of Color
   Tristimulus Colorimetry
   Specialized Colorimeters

3. **Interpretation of Data • 15**
   Presentation of Samples
   Interpretation
   Color Tolerances
   Blending of Colorants

4. **Regulation of Colorants • 23**
   Testing for Toxicity
   Acceptable Daily Intake
   European Union
   United States: food laws • Delaney Clause • sources of information

5. **FD&C Colorants • 33**
   Safety of Colorants
   Stability in Foods
   Lakes

6. Carotenoids • 43
Annatto
Saffron
Paprika
Tagetes
Lycopene
Miscellaneous Carotenoid Extracts
Synthetic Carotenoids
Health Aspects

7. Anthocyanins and Betalains • 55
Anthocyanins: chemical composition • commercial preparation • synthetic compounds • considerations in commercial applications • health effects
Betalains: chemistry • occurrence • commercial preparation • applications

8. Chlorophylls, Haems, Phycobilins, and Anthraquinones • 67
Chlorophylls: chemistry • preparation of colorants • applications
Haems
Phycobilins: chemistry • extraction • applications
Anthraquinones: cochineal and carmine • kermes • lac • alkannet

9. Turmeric, Carthamin, and *Monascus* • 77
Turmeric: properties • applications
Carthamin: properties and application
Monascus: chemistry • commercial production • applications

10. Caramel, Brown Polyphenols, and Iridoids • 83
Caramel: commercial preparation • chemistry • applications
Brown Polyphenols: cacao • tea
Iridoid Pigments: chemistry • production • applications

11. Miscellaneous Colorants • 89
Inorganic Colorants: titanium dioxide • Carbon Black • Ultramarine Blue • iron oxides • talc • zinc oxide • calcium carbonate • silver • silicon dioxide • others
Organic Colorants: fruit and vegetable extracts • riboflavin • corn endosperm oil • algae products • cottonseed products • shellac • octopus ink and squid ink

12. Baked Goods, Cereals, and Pet Foods • 97
Baked Goods and Cereals: FD&C colorants • natural colorants • shades
Pet Foods: water-soluble colorants • lakes • exempt colorants
Troubleshooting

13. Beverages and Dairy Products • 107

Beverages: FD&C colorants • natural colorants

Dairy Products and Spreads: butter • margarine • spreads • yogurt • fat-based coatings • retorted milk products

Troubleshooting

14. Confections • 117

Candy Starch Jellies and Candy Cream Centers: candy starch jellies • candy cream centers

Pan-Coated Candies: coating equipment • sugar panning • exempt colorants (natural colors)

Hard Candies (Boiled Sweets)

Other Products: direct-compression tablets • oil-based summer coatings • hard candy wafers • gum products

Troubleshooting

15. Special Topics • 125

Handling Colorants and Maintaining Color Quality: handling water-soluble colorants • preparing water-soluble colorants for use

Factors Affecting Colorant Quality

16. Future Prospects • 129

Appendix. Description of Colorants • 131

Glossary • 139

Index • 141

Overview

In This Chapter:
Adulteration of Foods
Appreciation of Color

The use of *color* in foods and cosmetics dates back to antiquity. Recorded history is replete with accounts of the use of *colorants* in everyday life. The art of making colored candy is depicted in paintings in Egyptian tombs as far back as 1500 B.C. Pliny the Elder described the artificial coloration of wines in 400 B.C. Spices and condiments were colored at least 500 years ago. Colorants in cosmetics were probably more widespread than those in foods or at least were better documented. Archaeologists have pointed out that Egyptian women used green copper ores as eye shadow as early as 5000 B.C. Henna was used to dye hair, carmine to redden lips, and kohl, an arsenic compound, to blacken eyebrows. It was common practice in India thousands of years ago to color faces yellow with saffron or to dye feet red with henna. The Romans put white lead and chalk on their faces and blue dyes on their hair and beards. More recently, in Britain, sugar was imported in the twelfth century in attractive red and violet colors, probably from the colorants madder and kermes for the reds and Tyrian purple for the violet colors.

Adulteration of Foods

Ancient history contains many examples of concern about food adulteration and the regulation of foods. This can be seen in the laws of Moses, the Roman Statutes, and documents down through the Middle Ages. For example, in 1266, the English Parliament prohibited a number of important food staples if they were so adulterated as to be "not wholesome for Man's Body." This unambiguous statement could form the basis of food regulation today (1).

Early applications of colorants were usually associated with religious festivals and public celebrations. If colorants were added to foods, their purpose was not to misrepresent the product. In the Middle Ages, the trade guilds policed their members and protected the quality of their products. However, with the evolution of the industrial revolution, with its massive social and population changes and much wider food distribution, quality safeguards were weakened and adulteration of foods became rampant. Wines and confectionery were particularly suspect. For example, Walford (2) described a recipe for bogus English claret as quoted in F. A. Filby's *The Innkeepers and Butlers Guide of 1805*:

Colorant, color—A *colorant* is a pigment used to color a product, as distinct from a *color*, which is used in this book in the physical sense, to indicate a hue such as red, green, blue, etc.

> Take six gallons of water, two gallons of cyder and eight pounds of Malaga raisons, bruised, put them all together and let them stand close covered in a warm place for a fortnight, stirring every other day very well. Then strain out the liquor into a clean cask and put to it a quart of barberries, a pint of juice of raspberries and a pint of juice of black cherries. Work it up with a little mustard seed and cover it with a piece of dough three or four days by the fireside and let it stand a week and bottle it off, and when it becomes fine and ripe it will become like common claret.

In 1820, Accum (3) published a book describing practices such as the coloring of tea leaves with verdigris (copper acetate), Gloucester cheese colored with red lead, pickles boiled with a penny to make them green, confectionery with added red lead and copper, and many other appalling practices. Hassall published a book on adulteration of foods in 1857 (4) that raised the public conscience, and then things began to change. Nearly every country passed laws dealing with food adulteration. Colorants were a major concern because they could be used to conceal inferior food and in some cases were actually dangerous.

Until the middle of the nineteenth century, the colorants used in cosmetics, drugs, and foods were of natural origin, from animals, plants, and minerals. After 1856, when Sir William Henry Perkin discovered the first synthetic organic dyestuff (mauve), the supply situation changed drastically. The German dyestuff industry synthesized a number of "coal tar" dyes, which apparently were very quickly utilized, as French wines were colored with fuchsine as early as 1860. In the United States, colorants were allowed to be added to butter in 1886 and to cheeses in 1896. By 1900, Americans were eating a wide variety of colored foods such as jellies, cordials, ketchup, butter, cheese, candy, ice cream, wine, noodles, sausage, confectionery, and baked goods.

Many of these practices constituted economic fraud but were not actually dangerous. However, some were. Marmion (5) described a classic horror story in 1860 in which a druggist gave a caterer copper arsenite to use in making a green pudding for a public dinner, and two people died. At the turn of the century, nearly 700 synthetic colorants were available, but very little control was exercised over the type and purity of the colorants offered for use in foods. Because of public concerns over these practices, American food manufacturers decided to police their own industry. One result was a list, published in 1899 by the National Confectioners Association containing 21 colorants that the association considered unfit for addition to foods.

However, it became apparent that some form of government control was needed, and, in 1900, the Bureau of Chemistry in the U.S. Department of Agriculture (USDA) was given funds to study the problem. A series of food inspection decisions (FIDs) resulted. A 1904 FID declared a food to be adulterated if it be colored, powdered or polished with intent to deceive or to make the article appear to be of better quality than it really is.

Another FID, in 1905, required the labeling of such products as artificially colored imitation chocolate. A 1906 FID concerned a coal tar dye considered to be unsafe for use in food and, in effect, stopped the importation of macaroni colored with Martius Yellow. At this time, the USDA launched a monumental effort to determine which colorants were safe for use in food and what restrictions should be placed on their use. This led to the regulations for synthetic colorants that are in use in the United States today (see Chapter 4).

Appreciation of Color

The enjoyment of food is associated with texture, flavor, odor, and color. Numerous studies have shown that food flavor and color are closely associated. Consumers expect certain foods to be associated with certain colors, and surprises are seldom appreciated. Goldenberg (6) described an experiment in the United Kingdom with canned garden peas, canned strawberries, and strawberry and raspberry jam. Without added color, which consumers in the United Kingdom apparently expected, the peas were greenish gray, the strawberries were a pale straw color, and the jams were a dull brown. Consumer reaction was swift and negative, and sales suffered accordingly. When the expected color was restored, sales slowly returned to former levels.

Nearly every food product has an acceptable range of color, which depends on a wide range of factors. These involve the ethnic origin of both the consumer and the food, age and sex of the consumer, physical surroundings at the time of viewing, health consciousness of the consumer, physical well-being, etc. Foods outside the normal range of acceptable color are rejected, sometimes rather emphatically. In one study, a dinner was composed of steak, French fries, and peas, which were abnormally colored but served under lighting conditions that masked the colors. During the meal, when normal lighting was restored, the panelists could see the blue steak, red peas, and green French fries. The reaction of the panelists was such that some of them reported feeling ill. Such extreme displays make interesting anecdotes, but they have little impact on everyday life. In some instances, unusual colors have even become accepted, such as green beer and green mashed potatoes on St. Patrick's Day.

Color may influence expectations in more subtle ways, particularly when the flavor of the food is close to the identifiable threshold. For example, Hall (7) prepared sherbets with lemon, lime, orange, grape, pineapple, and almond flavors. When the sherbets were mismatched in color, many panelists could not identify the flavors. For example, when the lime-flavored sherbet was colored green, 75% of the judges said it was lime, whereas when it was colored purple, as in grape, only 42% identified it correctly. In other research with orange beverages, the perceived intensity of flavor increased with increasing color.

Apparently, one can increase the perceived flavor level by increasing the appropriate color. This effect may be useful to food manufacturers, since colorants are usually cheaper than flavorings.

In other experiments, chocolate-flavored white ice cream and vanilla-flavored brown ice cream were presented to a trained panel. When asked to identify the flavor, the tasters said that the white ice cream was vanilla and the brown ice cream was chocolate. Similar tests with children using red and yellow jellies revealed that the red jellies were thought to be strawberry and the yellow ones lemon, regardless of the actual flavors. This type of experiment has been repeated many times, particularly at expositions and trade shows, and the results are predictable, based on consumer expectations, provided the flavor thresholds are close to the detectable limit.

Color is associated with the psychological aspects of various moods (8). Red is associated with power and possibly protection. Orange and yellow confer a sense of excitement, whereas blue is associated with pleasantness and security. Black has a strong association with hostility and dejection and is linked with death and disaster in Western countries. This association doesn't seem to faze the fashion industry, which considers black an elegant color, but it may have some influence on food. Few food products are packaged in black containers.

Clydesdale (9) reported that

> unquestionably color influences other sensory characteristics, and, in turn, food acceptability, choice, and preference. However, its effect seems to result from learned association rather than from inherent psychophysical characteristics . . . .

Numerous studies have shown that sugar solutions colored dark red are perceived to be sweeter than others of the same sucrose concentration in lighter colors or with distilled water. Similarly, red solutions are perceived to have a lower sucrose sweetness threshold and a higher pleasantness rating. Color seems to have little effect on the perception of saltiness, perhaps because saltiness is not associated with any single food in nature.

Color may influence acceptability in other ways because a complex relationship exists between color and acceptability as related to age (10). A group of researchers (11) tested the acceptability of 15 beverages differing in flavor, color, and sucrose levels, using a "student" panel aged 18–21 years and an "elderly" panel over 60. The two groups reacted quite differently. For overall acceptability, the older group reacted more positively to higher levels of color. Color had a significant inverse effect on sweetness at the highest level for both groups and a direct effect at the lowest level for the elderly group. It is obvious that the interaction of color with age is a complex phenomenon but one that could be very important in food formulation as our population grows older.

Food colorants are an important aspect of the formulation of foods. Walford (2) summarized the reasons for adding colorants to food today:

1) To give an attractive appearance by replacing the natural color destroyed during processing or likely to be destroyed by anticipated storage conditions.

2) To give color to those processed foods such as soft drinks, confectionery, and ice creams that otherwise would have little or no color.

3) To supplement the intensity of the natural color where this is perceived to be weak.

4) To ensure batch-to-batch uniformity when raw materials of different sources and varying color intensities have been used.

With today's emphasis on increasing food production for an ever-growing population and the necessity of making food appealing and acceptable, food colorants will remain a major concern. They already are an important part of the economy. Taylor (12) estimated that the worldwide market for colorants in 1994 was about 265 million U.S. dollars. Of this, natural colorants represented 65%, based on dollar volume. Pszczola (13) estimated the market for natural colorants in 1998 to be $939 million. The real market value can be a direct function of how one interprets or defines natural colorants.

References

1. Hutt, P. B. 1996. Approval of food additives in the United States: A bankrupt system. Food Technol. 50(3):118-128.
2. Walford, J. 1980. Historical development of food coloration. Chap. 1 in: *Development of Food Colours*, Vol. 1. J. Walford, Ed. Applied Science Publishers, London.
3. Accum, F. 1820. *A Treatise on the Adulteration of Food and Culinary Poisons*. London.
4. Hassall, A. H. 1857. *Food and Its Adulterants: Comprising the Reports of the Analytical Sanitary Commission of the Lancet 1857*. London.
5. Marmion, D. M. 1984. *Handbook of U.S. Colorants for Foods, Drugs and Cosmetics*, 2nd ed. Wiley-Interscience, New York.
6. Goldenberg, N. 1977. Why additives?—The safety of foods. In: *Colours—Do We Need Them?* Nutrition Foundation, Washington, DC.
7. Hall, R. L. 1958. Flavor study approach at McCormick & Co., Inc. In: *Flavor Research and Food Acceptance*. Arthur D. Little, Inc., Ed. Rheinhold Publishing Corp., New York.
8. Hutchings, J. 1996. Colour me beautiful. World Ingredients 1:28-29.
9. Clydesdale, F. M. 1993. Color as a factor in food choice. Crit. Rev. Food Sci. Nutr. 33:83-101.
10. Francis, F. J. 1995. Quality as affected by color. Food Qual. Pref. 6:149-155.
11. Phillipsen, D. H., Clydesdale, F. M., Griffin, R. W., and Stern, P. 1995. Comparison of sensory responses on an elderly population versus a college age population using a carbonated drink. J. Food Sci. 60:364-368.
12. Taylor, R. F. 1996 Natural food colorants as neutraceuticals: Markets and applications. *Proceedings of the Second International Symposium on Natural Colorants*. P. Hereld, Ed. The Hereld Organization, Hamden, CT.
13. Pszczola, D. E. 1998. Natural colors: Pigments of imagination. Food Technol. 52(6):70-82.

CHAPTER 2

Measurement of Color

The appreciation of color involves an effect on the human brain via a signal from the eyes. It should not be surprising to learn that a large body of science has evolved around the physiology of vision and the interpretation of visual signals from the human brain. Vision seems to be the top intelligence gatherer, since nearly one quarter of the brain's cerebral cortex is devoted to sight, much more than is devoted to the other senses.

The color of a food is not a physical characteristic such as particle size, melting point, or specific gravity. Rather it is one portion of the input signals to the human brain and results in the perception of one aspect of the food's appearance. Appearance may be influenced by a number of physical characteristics, such as particle size or texture, and a series of psychological perceptions, such as background color or arrangement in space. The human eye is a very sensitive organ—it can detect up to 10,000,000 different colors. Color as seen by the human eye is an interpretation by the brain of the character of light being reflected from or transmitted through an object. It is possible to define a food's color in a purely physical sense in terms of the physical attributes of the food, but this approach has serious limitations when we try to use color measurement as a quality control tool for food processing and merchandising. A more satisfying approach is to define color in a physical sense as objectively as possible and interpret the output in terms of how the human eye sees color.

In This Chapter:

Physiological Basis of Color

Visual Systems

Spectrophotometric Measurement of Color

Tristimulus Colorimetry

Specialized Colorimeters

Physiological Basis of Color

It is possible to estimate rigorously the physical stimuli received by the human eye, which eventually are interpreted as color, but unfortunately this is not true for physiological reactions (1). The initial stimuli by which we perceive color have been described, and the responses of the eye have been standardized (2–4). Briefly, the human eye has two types of sensitive cells in the retina, the *rods* and the *cones* (5). The rods are sensitive to lightness and darkness and the cones to color. Of the three types of cones within the retina, one is sensitive to red light, another to green, and the third to blue. This may be an oversimplification since individuals may have up to nine genes that encode for the cones and two types of red cones have already been discovered. Individuals may differ in the physiological response of

Rods—Long, slender bodies in the eye that respond to light and dark.

Cones—Conical receptors in the eye that respond to color.

the cones, but the variations must be small because color vision has been characterized and standardized in terms of the standard observer. The cones send signals to the brain, which interprets them in terms of opposing pairs. One pair is red-green and the other pair is yellow-blue. This is why we have individuals who are red-green or blue-yellow *color-blind*. There are no individuals who are red-yellow or green-blue color-blind.

The interpretation of the signals from the retina is a complex phenomenon influenced by a variety of physiological and psychological aspects. One aspect is color constancy; a sheet of white paper looks white in bright sunlight and also when under the green leaves of a tree with green light reflected onto it. The physical stimuli are different, but the brain knows that the paper should be white. Similarly a large expanse of color appears brighter than a small area. One needs only to paint a whole wall of a room a particular color and see how different it appears from the color of a small color chip in a paint store. There are many examples of this type of color interpretation by the human brain. The old adage "I believe what I see" makes interesting conversation but unfortunately is not always true, since it is a relatively simple matter to fool the eye. A classic demonstration shows a triangle with three right angles. This is obviously impossible, but it is only when we see a view from another angle that we realize that the sides of the triangle do not meet in space. In this situation, the human brain is not given sufficient information to make a correct judgment, but the brain does make some judgment, based on available information, which may or may not be correct.

Visual Systems

A number of excellent three-dimensional *color solids* for visual measurement of color have been developed. Perhaps the best known in America is the Munsell system (6), which contains 1,225 color chips arranged for convenient visual comparison. Each chip has a numerical designation that provides an unambiguous description of that color. This system is described in more detail in Chapter 3. The designation of colors of foods by visual comparison is very appealing since it is simple, convenient, and easy to understand. Many specialized color standards of paint, plastic, or glass are available, and a number of companies have adopted this approach for food quality purposes. For example, the U.S. Department of Agriculture's official grading system for canned tomato juice employs spinning disks of a specific Munsell designation to describe the color grades (7). Glass color standards are available for sugar products. Plastic color standards are available for a large number of products such as apple butter, peanut butter, peas, lima beans, orange juice, canned mushrooms, peaches, salmon, sauerkraut, and pimento.

The visual glass and plastic color standards have been very successful, particularly for grade standards, but they are available in a limited

Color blindness—The inability to distinguish between chromatic colors.

Color solid—A three-dimensional solid within which every point represents a specific color. The solid can be characterized either mathematically or physically.

number of colors. Painted paper chips, as in the Munsell system, are available in a much wider array of colors, but even these are limited. The chips are also fragile and may change color with use, particularly when exposed to direct sunlight. Visual standards may have other problems, since repeated visual judgments are tiring and sometimes tedious. Colors that fall between existing standards are often difficult to communicate to other individuals. Visual perception of color is affected not only by light but also by product texture. This is why instrumental methods of color measurement have been so appealing.

Visible spectrum—The wavelengths of the spectrum (usually defined as 380–680 nm) that can react with the rods and cones in the eye to send a signal to the brain.

Spectral color—A wavelength within the visible spectrum.

Spectrophotometric Measurement of Color

The development of instrumental methods of color measurement was based on reflection, or transmission, spectrophotometry (8). Physiologists developed the response of the cones in the human eye in terms of the *visible spectrum* and were able to do this in a manner easily reproducible in a laboratory today. Three projectors are required, each with a red, blue, or green filter in front of the lens (Fig. 2-1). Red, green, and blue light beams are focused on a screen such that they overlap over half a circle. The other half is illuminated by another projector or by spectrally pure light from a prism or grating. The observer can see both halves of the circle simultaneously. Each projector is equipped with a rheostat so that the amount of light from each of the red, green, or blue sources can be varied and the observer can determine the amounts of red, green, and blue required to match almost any *spectral color*. Therefore, we can define spectral color in terms of the amounts of red, green, and blue. We can set up a triangle with red, green, and blue stimuli at the corners (Fig. 2-2). Every point within the triangle represents a color and can be specified mathematically by the amount of red, green, and blue. Unfortunately, red, green, and blue are not particularly good stimuli to use since not all colors can be matched with them, so the early researchers chose to use X, Y, and Z as terms for the reference stimuli. These cannot be reproduced in the laboratory since they are only mathematical concepts. However, they were chosen for optimum mathematical convenience in constructing a color solid and specifying *color coordi-*

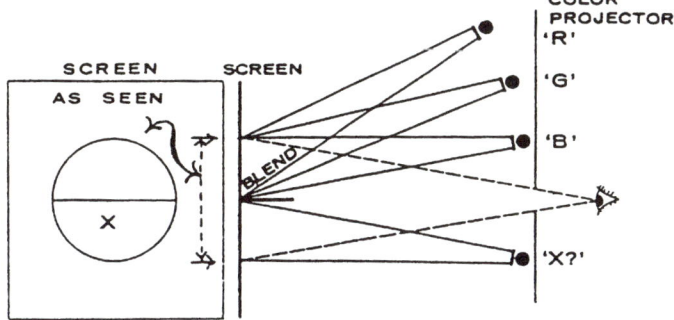

Fig. 2-1. Setup for determining response of the human eye to color. Three projectors are focused on the upper half of a circle on the screen. The color to be measured is projected on the lower half, and the eye can see both halves simultaneously. R = red, B = blue, G = green, X = color being matched. Reprinted, with permission, from (10).

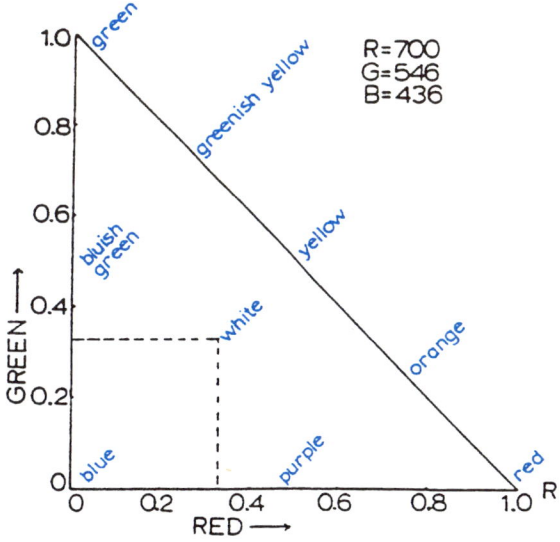

Fig. 2-2. Colors plotted on a red (R), green (G), and blue (B) color triangle. Reprinted, with permission, from (10).

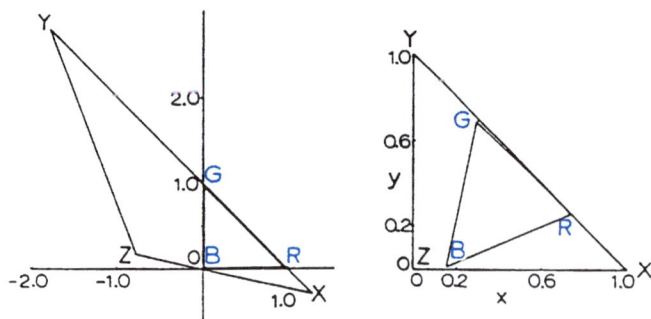

Fig. 2-3. The relative positions in space of the red-green-blue (RGB) and XYZ stimuli. Left, the *GRB* coordinates plotted as a right-angled triangle; right, the *XYZ* coordinates plotted as a right-angled triangle. Reprinted, with permission, from (10).

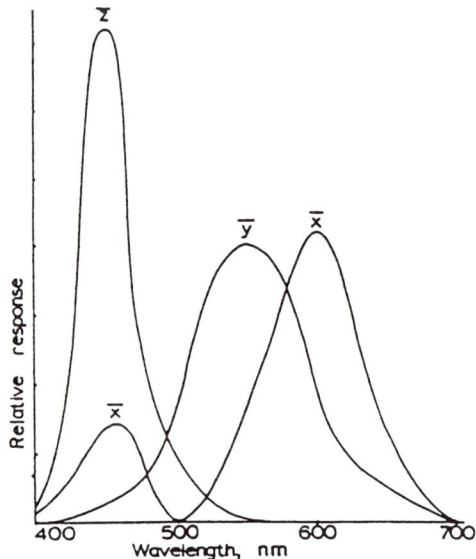

Fig. 2-4. Relationship between the response of the human eye, defined as the standard observer curves ($\bar{x}$, $\bar{y}$, $\bar{z}$), and the visible spectrum. From (7).

Color coordinates—Three coordinates within a color solid that locate a specific color in three-dimensional space.

nates. If one wants a crude visual reference, one can think of X as red, Y as green, and Z as blue. The relative positions in space for red, green, and blue and X, Y, and Z are shown in Figure 2-3.

If we take the red, green, and blue data for the spectral colors, transform them into X, Y, and Z coordinates, and plot the responses of the human cones for the visual spectrum (Fig. 2-4) against wavelength, we have the response of the human eye to color. These curves were standardized in 1932 and were called the CIE $\bar{x}$ $\bar{y}$ $\bar{z}$ standard observer curves. (The initials stand for "Commission Internationale d'Eclairage.")

With the data in Figure 2-4, it is mathematically simple to calculate the color from a reflection or a transmission spectrum, as illustrated in Figure 2-5. The sample spectrum is multiplied by the spectrum of the light source, and the area under the resultant curve is integrated in terms of the $\bar{x}$ $\bar{y}$ $\bar{z}$ curves of Figure 2-4. The process can be described mathematically by the integral equations

$$X = \int_{380}^{750} RE\bar{x}\, dx$$

$$Y = \int_{380}^{750} RE\bar{y}\, dy$$

$$Z = \int_{380}^{750} RE\bar{z}\, dz$$

where R = the sample spectrum, E = the source spectrum, and $\bar{x}$, $\bar{y}$, and $\bar{z}$ = the standard observer curves. The *X, Y, Z* data are usually plotted as *x, y, z* coordinates, where

$$x = X / (X+Y+Z)$$

$$y = Y / (X+Y+Z)$$

$$z = Z / (X+Y+Z)$$

The visible spectrum plotted on an *x, y* diagram is shown in Figure 2-6. The color solid is actually a solid, not a plane, with the lightness function perpendicular to the plane of the paper. Figure 2-6 also illustrates another popular way of presenting color data. The point of intersection of a line from the coordinates of white light (x = 0.333, y = 0.333) through a sample point to the edge of the solid is the dominant wavelength of the sample point. The relative distance of the sample point from white light is the purity of the color.

Early spectrophotometers provided a reflection or transmission spectrum, with manual or mechanical (and later, electronic) calcula-

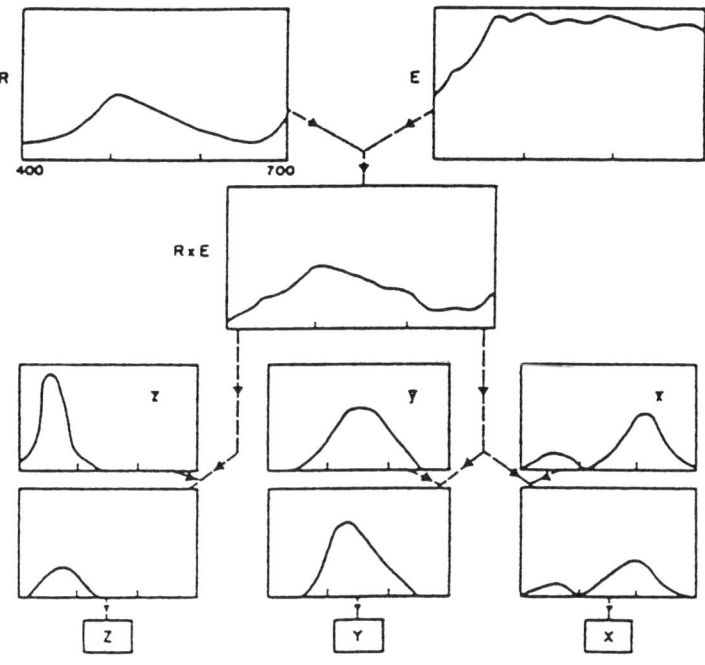

Fig. 2-5. Derivation of XYZ data from a spectrophotometric reflection or transmission curve. R = sample spectrum, E = source spectrum. From (7).

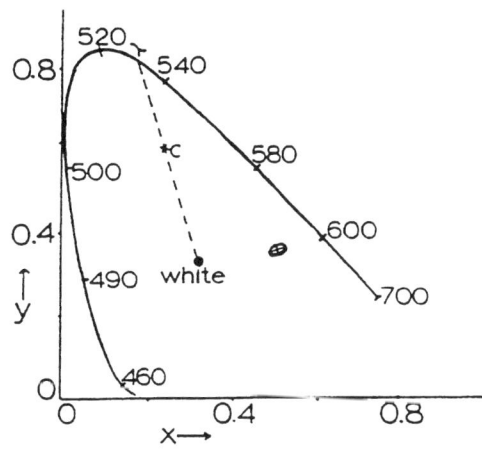

Fig. 2-6. The spectrum colors plotted on x, y coordinates. Reprinted, with permission, from (10).

tion of XYZ data. However, the early instruments were complicated and expensive, and this facilitated the development of *tristimulus* colorimeters.

Tristimulus Colorimetry

Definition of the standard observer curves led to the development of colorimeters designed to duplicate the response of the human eye. The concept is very simple (Fig. 2-7). One needs a light source, three glass filters with transmittance spectra that duplicate the *X*, *Y*, and *Z* curves, and a photocell. With this equipment, we can get an *XYZ* reading that represents the color of the sample. All tristimulus colorimeters available today depend on this principle, although they have individual refinements in photocell response, sensitivity, stability, and reproducibility. A number of color solids have been used in the past, but some standardization is developing. Three systems seem to be gaining in popularity. One is the XYZ system. Another is the Judd-Hunter L a b system, in which *L* is lightness or darkness, +*a* is redness, –*a* is greenness, +*b* is yellowness, and –*b* is blueness. A third is the L* a* b* system, similar in concept to the L a b system (9) and usually called CIELAB. Later the L* C* H* system was developed and usually called CIELCH. This system was developed primarily for color tolerances and is described more fully in the section on that subject.

Tristimulus—Describing something that reacts to the three values (red, green, and blue) that make up color.

Specialized Colorimeters

The success of the tristimulus colorimeters led to a great expansion in research on color in foods because data could be obtained very quickly with relatively inexpensive instruments. At roughly the same time, statistical quality control concepts for production control were being developed. Unfortunately, most statistical quality control charts were either one dimensional, or at best two dimensional, but color data was three dimensional. Requests emerged for color data to be reduced to one or two dimensions, and a series of approaches resulted (7). Specialized colorimeters were developed to provide a one-dimensional readout for a given food product. For instance, the tomato colorimeter measures raw and canned tomato juice, and the citrus colorimeter measures the color of citrus fruit. Specialized instruments were developed for honey, sugar, tea, apples, cranberries, salmon, wine, and the internal color of pork and beef (10). The proliferation of specialized instruments led to some dissatisfaction among suppliers, who did not want a roomful of instruments. The current trend is to use a general-purpose instrument to collect the data and then get a readout in any required unit simply by adding a simple computer circuit. Modern advances in computers have made calculation of color scales so economical that the computational labor is simply not a factor. Since the cost of computational labor led to the development of tristimulus colorimeters originally, we seem to have come full circle.

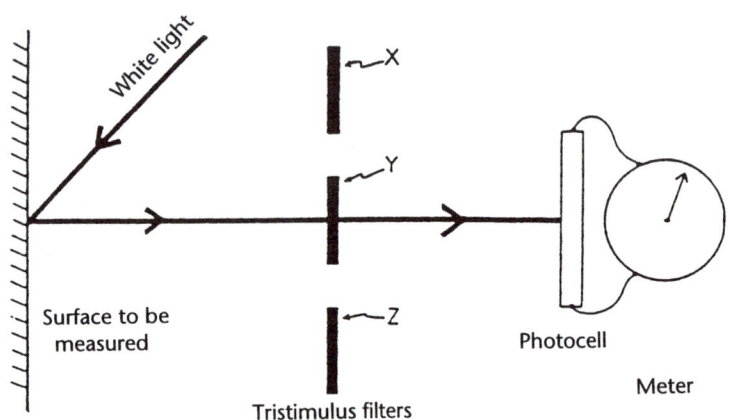

Fig. 2-7. Diagram of a simple tristimulus colorimeter. From (7).

References

1. Dember, W. N., and Warm, J. S. 1976. *Psychology of Perception*, 2nd ed. Holt, Rinehart and Winston, New York.
2. Wright, W. D. 1969. *The Measurement of Color*. Van Nostrand-Reinhold Co., New York.
3. Hutchings, J. B. 1994. *Food Colour and Appearance*. Blackie Academic & Professional, Glasgow, Scotland.
4. Kuehni, R. G. 1997. *Color: An Introduction to Practice and Principles*. John Wiley & Sons, New York.
5. Boynton, R. M. 1979. *Human Color Vision*. Holt, Rinehart and Winston, New York.
6. Anonymous. 1963. *Munsell Book of Color*. MacBeth, Newburgh, NY.
7. Francis, F. J., and Clydesdale, F. M. 1975. *Food Colorimetry: Theory and Applications*. AVI Publishing Co., Westport, CT.

8. Billmeyer, F. W., and Saltzman, M. 1981. *Principles of Color Technology*, 2nd ed. John Wiley & Sons, New York.
9. Hunter, R. S., and Harold, R. W. 1987. *The Measurement of Appearance*, 2nd ed. John Wiley & Sons, New York.
10. Francis, F. J. 1991. Color measurement and interpretation. In: *Instrumental Methods for Quality Assurance in Foods*. D. Y.-C. Fung, Ed. Marcel Dekker, New York.

CHAPTER 3

Interpretation of Data

For many years, workers in the color area had to cope with at least a dozen color systems, including both visual and instrumental approaches. Today, five systems are in general use—one visual and four instrumental. Many visual systems exist, but the one most commonly used in the United States is the Munsell color solid (1). The instrumental systems are the CIE XYZ, the Judd-Hunter L a b, the CIELAB L* a* b*, and the CIELCH L* C* H*. The visual systems are certainly the simplest and require little interpretation. The instrumental systems do require interpretation because they are rigorous and reproducible for ideal samples such as those that are completely opaque or perfectly transparent with no light scattering. Unfortunately, very few foods fit either category since most foods both transmit and reflect light.

In This Chapter:
Presentation of Samples
Interpretation
Color Tolerances
Blending of Colorants

Presentation of Samples

The Munsell color solid is organized in three dimensions (Fig. 3-1). An analogy to the structure of a citrus orange may be appropriate, where the Munsell *value* (*V*), lightness or darkness, corresponds to the

Value—The coordinate in the Munsell System that corresponds to the lightness-darkness scale.

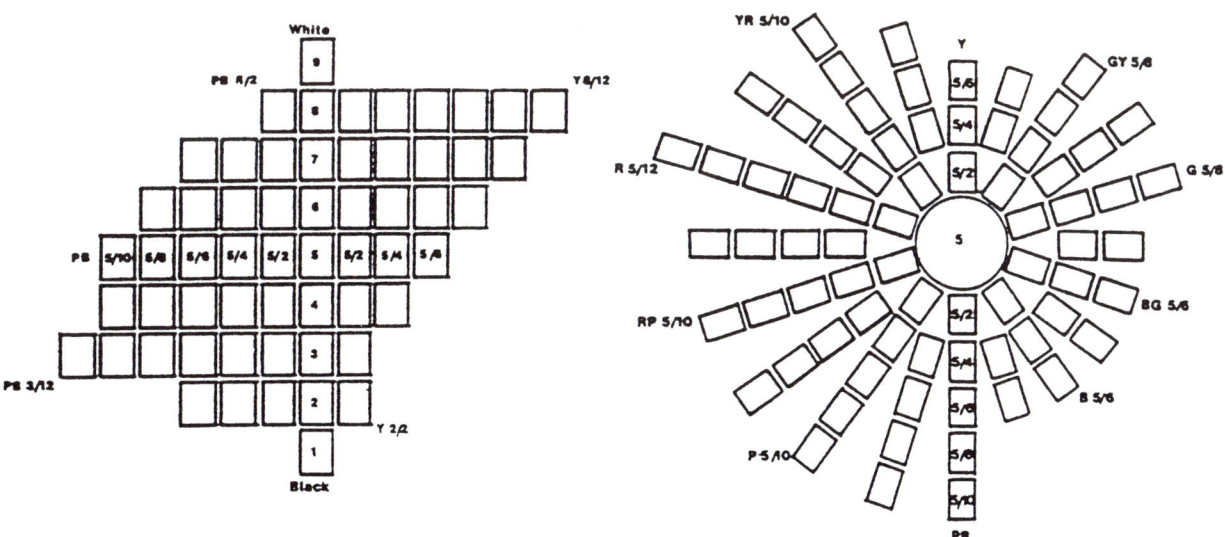

Fig. 3-1. The Munsell color solid, two views. Y = yellow, G = green, B = blue, P = purple, R = red. From (2).

Hue—Gradation of color. Also the coordinate in the Munsell System that corresponds to the red-green-blue scale.

Chroma—The coordinate in the Munsell System that corresponds to the color intensity scale.

Munsell Book of Color—A collection of color chips organized in a three-dimensional solid.

core of the fruit; Munsell *hue* (*H*), such as red, yellow, green, etc., corresponds to each section of the fruit; and Munsell *chroma* (*C*) relates to the distance from the core to the outside edge of the fruit. The *Munsell Book of Color* usually comes in a form comparable to a series of vertical slices through the fruit. Each slice shows a series of chips varying in *V* and *C* with a constant *H*.

For comparison with a food sample, a gray mask with an opening the size of one chip is placed over the sheet of chips and the one exposed chip is compared directly to the food sample. The mask, which minimizes the effect on the viewer of the surrounding chips, can be moved over the sheet of color chips until a color match with the food sample is found. If the food sample is much larger than the colored chip, finding a match can be awkward. An alternative is to purchase larger chips, which is feasible for specific situations. All of the judgments must be made under controlled lighting.

With the instrumental systems, the usual practice is to measure color by reflection from the surface of the sample or by transmittance through the sample. For reflectance, the simplest procedure is to fill a cuvette with the sample and measure the light reflected from the bottom of the cuvette. The cuvette should be filled to infinite thickness such that adding more product to the cuvette does not change the reading. The same cuvette can be used to measure transparent samples by placing a ring, usually 1 cm in depth, inside the cuvette and placing a white plastic disk on the ring. Since transparent food samples are usually liquid, the cell or cuvette can be filled above the level of the disk. In effect, the light enters the cell from the bottom, is transmitted through the sample, and is reflected off the white disk back through the sample and into the measuring port of the colorimeter. The effective path length of measurement is 2 cm. Alternatively, the sample can be placed in a vertical cell, usually 1 cm in light-path length, and placed in the light beam of a transmission attachment for a colorimeter.

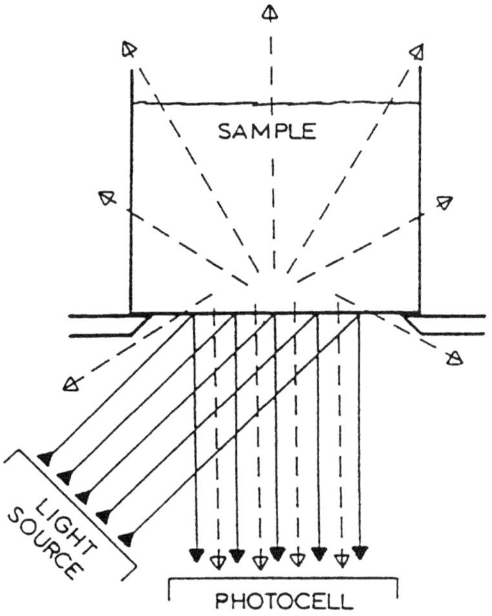

Figure 3-2 shows a typical presentation for a food sample for measurement by reflection. Light enters from the source beam below, and some is reflected from the glass surface into the measuring port of the instrument. Some of the light enters the sample and is scattered. Part of the scattered light is reflected back into the measuring port, but some escapes through the sides of the cell. Thus, light entering the sample is usually partially scattered and partially absorbed, and the instrument measures the light that emerges back into the measuring port. It is obvious that the instrument response is empirical, but it is usually reproducible for a given situation. Turbid samples may show less reproducibility since the turbidity may not be uniform or reproducible. In such a situation, the reproducibility can be increased by filtering out the particles that scatter the light. This is seldom an advantage since

Fig. 3-2. Interaction of a light source with a food sample to produce a signal. Reprinted, with permission, from (1).

the usual desire is to measure samples as a consumer sees them. The problems with light scattering and absorption can be handled by the Kubelka-Munk equations (2), but these have not been accepted to any extent by the food industry. The usefulness of the data depends on the ingenuity of the operator in presenting samples to the instrument in a manner that will yield useful information. For consumer acceptance studies, the data should reflect consumer judgment of the product, but for production, quality control, stability, etc., the samples could be measured in any appropriate manner.

Interpretation

All color solids are three-dimensional, and this causes some difficulties since we simply do not think in three dimensions. Perhaps one way to visualize this is to realize that when one is asked to measure color, one is really being asked to locate the three coordinates in space. This identifies a color unambiguously, since a given point can be one color and only one color. A point in space can be visualized more easily with some systems than with others. For example, with the CIE XYZ solid, it is difficult for even experienced workers to visualize the color when given the XYZ values. This was part of the thinking when Richard Hunter developed the Judd-Hunter solid (Fig. 3-3). He had a number of things in mind, but three were more important than the others. First, he wanted to develop a solid with equal distances in each of three dimensions equal to the same degree of visual responses of the eye. In other words, he wanted an equidimensional solid. We now know that this is impossible, but the L a b solid comes a lot closer than the XYZ solid, and the L^* a^* b^* solid is a further improvement. Second, he wanted a color solid in which one could easily visualize the color. This is certainly true with the L a b solid. Third, he wanted a solid that would be related to the physiological basis for color vision, i.e., based on the opposing pairs theory. Most likely the ease of visualizing the color from the color coordinates was one of the reasons for rapid adoption of this system in the food field.

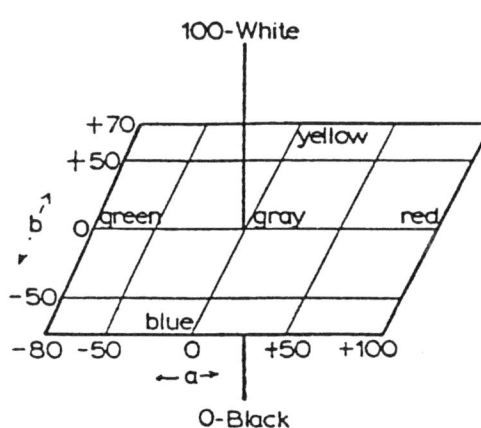

Fig. 3-3. The Judd-Hunter color solid. From (2).

It is likely that changes in storage, processing, ingredients, etc., result in color changes in all three color parameters, yet one or two may be more important. Perhaps one or two parameters can be discounted in practice, but this is a decision to be made in each application. People who present color data have a tendency to report each of the three parameters and do an analysis of variance on each parameter. But this assumes that all three are independent variables, which is simply not true. In the Judd-Hunter solid, a and b are related to each other and both are dependent on L. One popular alternative is to report data in the easily understandable L value and a measure of hue termed "theta" (Fig. 3-4). Theta (θ) can be defined as the angle

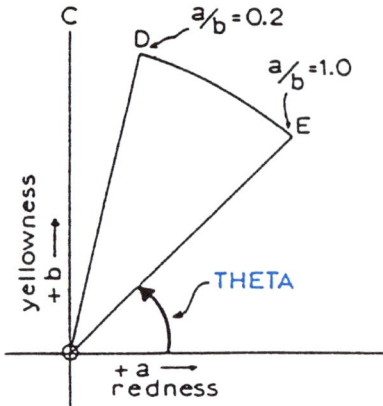

Fig. 3-4. Derivation of the hue function, theta, from an L a b diagram. Theta is the positive angle from 0 to 360°. Reprinted, with permission, from (3).

that a line joining a point in space with the origin makes with the horizontal axis. In Figure 3-4, if the value θ = 0, it would be a bluish red. Orange, yellow, green, and blue would be represented by θ = 45, 90, 180, and 270, respectively. Theta is obviously related to both *a* and *b*, and equal values of θ could differ in chroma. Chroma can be calculated as the square root of $a^2 + b^2$; it describes the distance from the sample point to the gray central axis. These calculated chroma values have been used very little in the food industry because chroma value is related to *L* and usually does not provide any useful additional data. The value θ can also be represented by the letter *h* (from the word *hue*), calculated in exactly the same way. One useful standardization would be to refer to the hue angle as θ when calculated from L a b data and as *h* when calculated from L* a* b* data. It is appropriate to use L, θ, or *h* in an analysis of variance. The simpler function *a/b* has been used in the past but is seldom appropriate. The *a/b* value is a tangential function, and as θ goes from 0 to 90, *a/b* goes from infinity to 1 to 0. Values approaching infinity create problems for analyses of variance (1).

One way to judge the importance of each parameter is to calculate correlation coefficients between, say, the visual judgments used in grading and instrument readings. For example, for canned tomato juice, a correlation between the visual judgment of graders and Hunter Rd (a scale related to *L* but no longer used) was –0.542. Correlations between the grade and Hunter *a*, *b*, *a·b*, and Rd *a·b* were +0.635, –0.652, +0.903, and +0.904, respectively. Apparently the judges were looking at both *a* and *b*, and the Rd value provided very little additional information. A similar experiment with lima beans indicated that only the *L* value was important, whereas with apple sauce all three values were required. However, the decision to discount one or two color parameters should be a conscious decision made with the realization that some information is being discarded.

Other approaches to the reduction of three parameters to one or two have involved the development of specialized scales and possibly their association with specialized instruments. One example is tomato juice color. For raw tomato juice,

$$TC = 2{,}000 \cos \theta / L$$

where TC is tomato color, cos q = $a/(a2 + b2)^{1/2}$. For processed tomato juice color, TC = (bL)/a.

A number of unidimensional scales have been developed, and there probably will be more because of their ease in computation and their usefulness in quality control procedures.

Color Tolerances

The description of color for purchase specifications or consumer acceptance of any commodity involves the concept of color tolerances. The color desired is located in color space, and allowable toler-

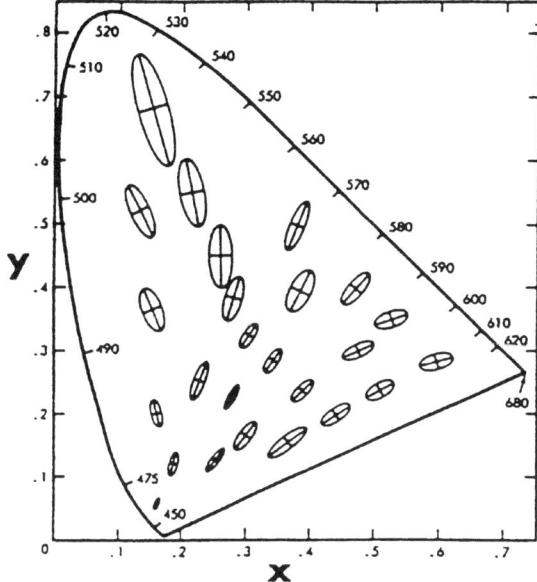

Fig. 3-5. Ellipses in color space representing the sensitivity of the human eye to various colors. From (2).

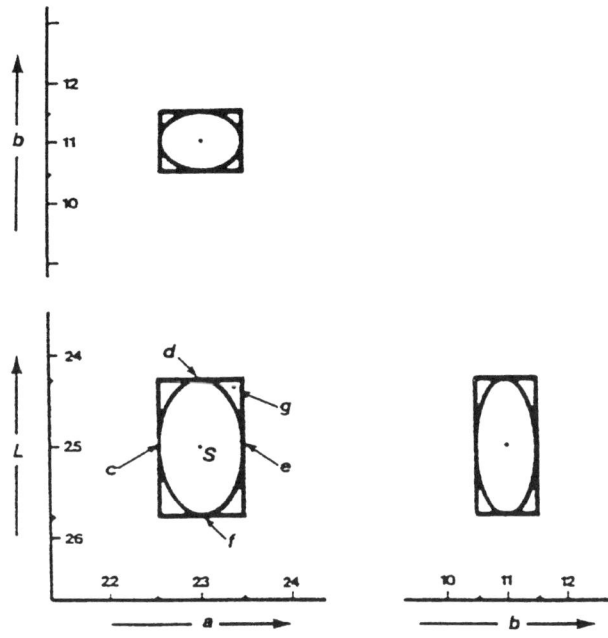

Fig. 3-6. Three-dimensional color tolerances in the L a b system. S = desired color of sample. Points c–f = points within the tolerance when an ellipse is used. Point g = point within the tolerance only when a rectangle is used. From (2).

ances are specified in one, two, or three dimensions. Some interesting specifications have been developed for a variety of purposes. For example, electrical wires are color coded because it is important that electricians be able to identify the proper wires. A color tolerance system was developed based on the Munsell System that gives the target color for the wire and acceptable tolerances in hue, value, and chroma. This involves a total of seven chips to specify the tolerance for each wire color. Similar situations exist with food products, but they may not always be three-dimensional, as described previously.

Color tolerances can be plotted in all color solids, but, unfortunately, it is not possible to specify a tolerance that is equally acceptable in all portions of the color solid. The reason for this is that the eye is much more sensitive to some colors. This is shown in Figure 3-5. The ellipses in the green area are much larger than in the blue or the red area, indicating a much greater sensitivity. Color tolerances may be plotted in three dimensions as shown in Figure 3-6. An ellipse is preferable to a rectangle for color tolerances in view of the sensitivity of the human eye. In Figure 3-6, if the point in the center of each area is the desired color (S), the points c, d, e, and f would be within the tolerance when an ellipse was used. The point g would be within the tolerance using a rectangle and outside using an ellipse. The L* C* H* system was developed as a more convenient way to handle tolerances. It is a polar system with the $L*$ the same as in the L* a* b* system. The $C*$ represents the vector distance from the center of the color space to the measured color and is a measure of chroma. $H*$ is a measure of hue cal-

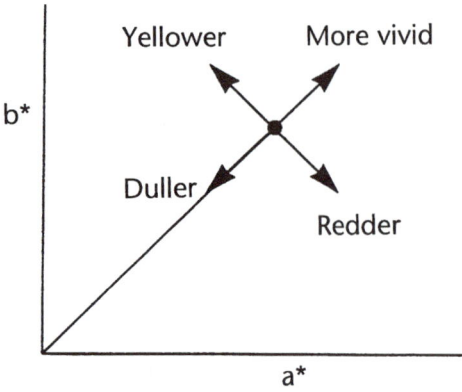

Fig. 3-7. Illustration of the visual effect of hue and chroma. The yellow-red axis is a measure of hue; the dull-vivid axis measures chroma.

Primary colors—Any three colors that, when mixed in suitable proportions, produce any color. Primary colors may work by subtraction of light (e.g., red, blue, and yellow) or by addition of light (e.g., red, green, and blue).

Additive colorimetry—Analysis of colors produced by the combination of three or more primary colors (e.g., red, green, and blue) or by the combination of spectral colors.

Pigment—A colored chemical compound.

Subtractive colorimetry—Analysis of colors produced by subtracting portions of the visible spectrum from white light.

culated in the same manner as θ. Figure 3-7 illustrates the difference between hue and chroma.

The L* C* H* polar system led to the development of the CMC color tolerancing system, which mathematically defines an ellipsoid around a standard color with the semi-axes corresponding to hue, chroma, and lightness. The ellipsoid, representing the volume of acceptance, automatically varies in size and shape depending on its position in color space. It also allows one to compensate for changes in sensitivity to L*. The eye generally has greater acceptance for shifts in the lightness dimension than for shifts in hue or chroma. The tolerance ratio of lightness to hue or chroma is usually about 2:1. Figure 3-8 shows a CMC ellipsoid with a ratio of lightness to hue or chroma (l:h/c) weighted at 2:1 and another weighted at 1.4:1. Usually the amount of color difference considered acceptable is defined in a single commercial factor (cf). The cf can be varied to make the ellipsoid as large or small as necessary. In many food applications, the degree of accuracy implied by the question of whether the color is within the rectangle but outside of the ellipsoid may not be necessary. The decision as to whether one, two, or three color parameters are necessary for color specifications is, of course, dependent on the use of the product.

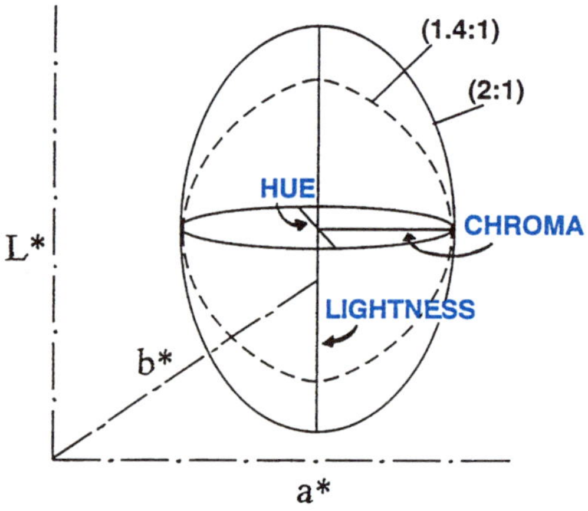

Fig. 3-8. A CMC ellipsoid with a l:h/c ratio of 2:1. The dotted line shows the effect of changing the l:h/c ratio to 1.4:1.

Blending of Colorants

It is unlikely that colorants will be available to produce the exact color desired for some products; therefore, formulators must blend different colorants. It is theoretically possible to create a desired color if three appropriate *primary colors*, such as red, green, and blue, are available, but, in practice, a wider range of colorants may be employed.

The mixing of colored lights to produce an array of colors was described in Chapter 2. This is known as *additive colorimetry*, in which a mixture of red, green, and blue light produces white light. The mixing of *pigments*, known as *subtractive colorimetry*, produces the opposite effect—a mixture of red, green, and blue produces a black color. The reason for this is that each pigment subtracts from white light a portion of the spectrum peculiar to that pigment, and if the whole visible spectrum is subtracted, a black color results. The recent introduction of electronic reconstruction in the photographic industry has produced a very sophisticated application of additive colorimetry. For example, the fashion industry theoretically has 16 million colors available for retouching photographs, but obviously some of them are below the threshold of visual discrimination of the human eye.

The laws of subtractive colorimetry are well known to color formulators. For example, in the auto repair business, repair shops are frequently called upon to repaint a fender or a door. A garage simply cannot stock all the colors required, so a simpler system is used. The mechanic looks up the color code for the car in question, which tells the type and amount of each primary paint required to match the color of the auto. The mechanic then makes up a batch and compares it with the auto's color. It probably won't match the first time, but, using personal judgment, the mechanic adds primary paint to make a match. Some formulators are amazingly good at this and seldom have to make more than two or three additions, at least for newer cars. The result is a paint repair job that is below the visual threshold for the difference between the color of the painted fender and the rest of the car.

The paint, plastic, and textile industries have developed the science of blending colorants to a very sophisticated degree. For example, a designer calls for a particular color to be mass produced for this year's designs and makes up a sample. A manufacturer makes a reflection curve of the sample using a spectrophotometer and enters the data into an instrument called COMIC ("color mixture computer"). This instrument, which has a library of reflection curves of available colorants and base materials, specifies what and how much of a given selection of colorants is required to match the sample. A batch of colorant is made up and applied to the base material, and a second curve is obtained. The original and second curves are entered into a second instrument called CODIC ("color difference computer"), which tells the operator how to adjust the colorant mixture to match the desired color. The system is sufficiently sophisticated that a third or fourth adjustment is seldom necessary.

The same degree of colorant sophistication could be used by the food industry, but it usually isn't, for several reasons. Possibly the most important is that the coloring of foods does not demand the same precision as that demanded by, say, the armed forces, which wants the trousers to match the tunic of a uniform. Also, variations in the physical state of food commodities and the chemical interactions of the colorants are broader. Possibly the breadth of applications does not merit the investment in the capital necessary to program instruments such as the COMIC and CODIC. The bottom line is that most colorant blending for foods is done visually, but this does

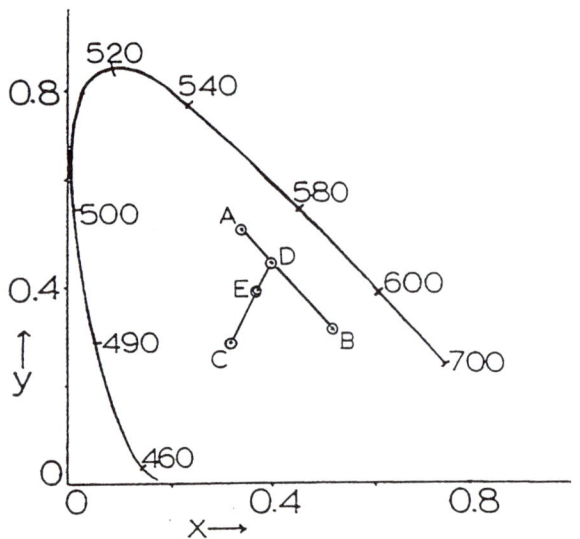

Fig. 3-9. Diagram showing blending of colorants (A–C) to produce a desired color (D or E).

FD&C colorants—Colorants certified by FDA as safe for use in food, drugs, and cosmetics. Also called certified colorants.

not mean that color diagrams cannot be utilized to improve visual judgments. Any desired color can be plotted on an x,y or an a,b diagram together with a given mixture of colorants. The position of the points helps the operator decide what and how much of a colorant should be added to make a closer match. For example, Figure 3-9 shows three colorants, A, B, and C, plotted on a CIE xy diagram. In order to match color D, a mixture of A and B would be required in the ratio of AD parts of B and BD parts of A. The match for color E would require C and D in the ratio CE parts of D and DE parts of C. This calculation refers only to hue, and a formulator may also have to adjust for lightness or darkness by adding white or black. In practice, the magnitude of the color adjustments is much less than that indicated in Figure 3-9. After several trials, formulators develop a surprising ability to judge how to vary a given blend to obtain a desired color.

One precaution is in order. Different colorants may react differently in products with varying particle size, acidity, oxidation-reduction potential, and other physical and chemical attributes. The color of each blend should be measured in the actual food for which it is intended. A series of blends to produce specific colors is shown in Chapter 5 (Table 5-10).

The color is an indication of the influence of the colorants in the food on the human eye, but it is not a substitute for chemical analysis of the pigment content. For reasons of product formulation, pigment stability and degradation, or component interaction, it may be desirable to measure the pigment content. A description of the methods of analysis for each colorant is beyond the scope of this book. For detailed descriptions of the analysis of the *FD&C colorants*, the reader is referred to *Handbook of U.S. Colorants* by Marmion (4).

References

1. Francis, F. J. 1995. Colorimetric properties of foods. In: *Engineering Properties of Foods*. M. A. Rao and S. S. H. Rizvi, Eds. Marcel Dekker, New York.
2. Francis, F. J., and Clydesdale, F. M. 1975. *Food Colorimetry: Theory and Applications*. AVI Publishing Co., Westport, CT.
3. Francis, F. J. 1991. Color measurement and interpretation. In: *Instrumental Methods for Quality Assurance in Foods*. D. Y.-C. Fung, Ed. Marcel Dekker, New York.
4. Marmion, D. M. 1991. *Handbook of U.S. Colorants*, 3rd ed. John Wiley & Sons, New York.

CHAPTER 4

Regulation of Colorants

The development of the regulatory aspects of food colorants is one small portion of the much bigger picture of regulation of food additives and food quality in general. One of the driving forces for food regulation down through the years has been food safety. This aspect was brought to public attention, sometimes in rather stormy fashion, by books such as *The Jungle* by Upton Sinclair in 1906 and more recently by the media. Public laws to minimize obviously unpalatable practices concerned with food sources, processing, handling, ingredients, etc., date back only about 150 years. Much of the criticism has been directed toward the artificial colorants and very little toward the naturally occurring colorants. Perhaps this is partially due to the naive belief that humans have somehow become conditioned to tolerate certain compounds after a long history of use. It is unlikely that 5,000 years is enough to create significant genetic changes in humans. It is true, however, that history has allowed us to identify acute hazards and eliminate them from our diet. Long-term or lifetime exposures are another story. Regardless, much of the regulatory activity has centered around food safety.

In This Chapter:

Testing for Toxicity

Acceptable Daily Intake

European Union

United States
Food Laws
Delaney Clause
Sources of Information

Testing for Toxicity

Regulation of the safety of food colorants is similar in principle around the world, but individual countries differ in both protocol and interpretation. In general, safety testing for all compounds depends on animal testing. Animals, usually rodents, are fed increasing amounts of the compound under test, and the effects are determined. A simplified diagram of the relationship between dosage and effect is shown in Figure 4-1. In Figure 4-1A, the increasing dosage, within reasonable limits, has no effect; examples would be sugar or starch. In 4-1B, every level of dosage creates a response. The classical example of this situation is radiation; few, if any, compounds fall into this category. In 1C, no response is seen until the body's defense mechanisms are overwhelmed, but then increased dosage produces a response. Nearly all chemicals fall

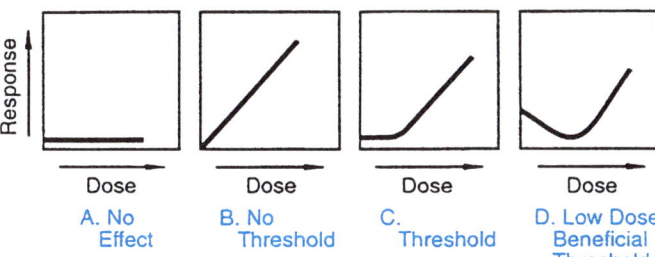

Fig. 4-1. Four simplified responses to a food additive. Adapted from (1).

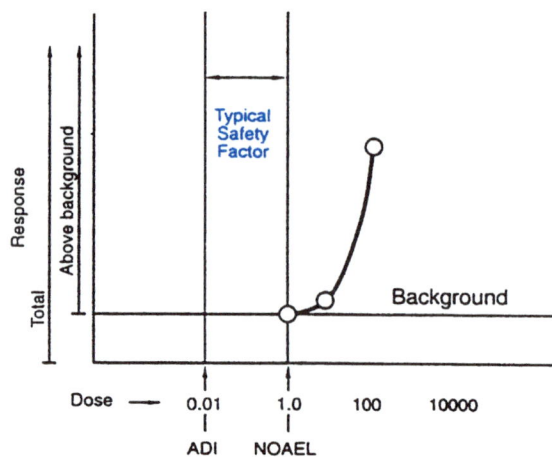

Fig. 4-2. A simplified dose response experiment with an animal model. ADI = acceptable daily intake. NOAEL = no observed adverse effect level (comparable to the no observed effect level, NOEL). Adapted from (5).

Acceptable daily intake (ADI)—The amount of a specific food additive thought to be the maximum level that should be ingested on a daily basis, as determined by experts such as those at WHO/FAO. Usually includes a 100-fold safety factor, but the safety factor may range from 10 to 5,000. Usually expressed as milligrams of the test substance per kilogram of body weight per day, based on a hypothetical average person weighing 70 kg.

into this category. In 1D, the immediate response is harmful. This is followed by a beneficial optimum and then by a harmful response. Oxygen, selenium, and vitamins A and D are in this category.

In Figure 4-2, a simplified experiment for chemicals in category C is shown. First, the total response (i.e., adverse response) is determined. This can be important because a high total response is indicative of problems with the experiment. Next the background level of response is determined, because most experiments have a small random response not related to the treatment. The next step is to determine the dosage level at which no effect can be observed. This is called the no observed effect level (NOEL) or the no observed adverse effect level (NOAEL). The NOEL is divided by 100 to obtain the *acceptable daily intake* (ADI). This 100-fold safety factor is derived from a factor of 10 to change animal data to human conditions and another 10 to account for human variability. The 100-fold safety factor is accepted worldwide and works well for many chemicals. However, if there are nonreversible immunologic or reproductive effects, the safety factor may be 1,000 or more. With carcinogens, the safety factor may be as high as 5,000; carcinogens create special problems because the time span for humans to experience observable effects may be a generation or more. In any event, many uncertainties lie in the determination of the NOEL, the extrapolation of high doses to low doses, and the interpretation for humans. A number of statistical methods are available to extrapolate from high to low doses, and they sometimes vary in estimates of risk by a factor of 200,000. One extreme example is dioxin, for which seven methods of calculating risks produced a 14-fold difference. When the uncertainties due to the interpretation for humans are added, the degree of uncertainty rises even more. This makes it possible for organizations to interpret the same data according to a "best case" or a "worst case" scenario according to their own agenda and to be statistically correct either way.

One way to reduce the level of uncertainty is to get a better understanding of the mechanism of action, thereby enabling the researcher to choose a more appropriate mathematical model to extrapolate from high to low doses rather than simply using a linear model. Linear models are often used by default because of insufficient information. A better understanding may also allow for the choice of a more appropriate animal model. For example, the carcinogenic effect of saccharin in rats is mediated through a protein that occurs in rat urine but not in humans. Rat data on saccharin are clearly unsuited for application to humans.

The "worst case" scenario in the classification of carcinogens has been criticized in a number of ways. First, use of the maximum tolerated dose has been accused of promoting tumors. This occurs because the injured tissue may have greater cell division, which in itself promotes more mutations, some of which may become carcinogenic. Second, many experiments use the $B_6C_3F_1$ strain of mice, which is very sensitive to tumor production. Third, many classifications do not include the degree of potency. All of these concerns increase the degree of uncertainty.

The use of a linear extrapolation when the response curve is nonlinear usually increases the response. Figure 4-3 shows a simplified case of extrapolation from one type of curve to the other. The curves are being used to find the "virtually safe dose" (VSD), a statistical interpretation that seems to be gaining favor, possibly because it is easy to understand. VSD is the dose at which the response curve crosses the acceptable risk level. In Figure 4-3, the VSD of the nonlinear model is virtually double that of the linear model. Again this points out the desirability of understanding the shape of the response curve.

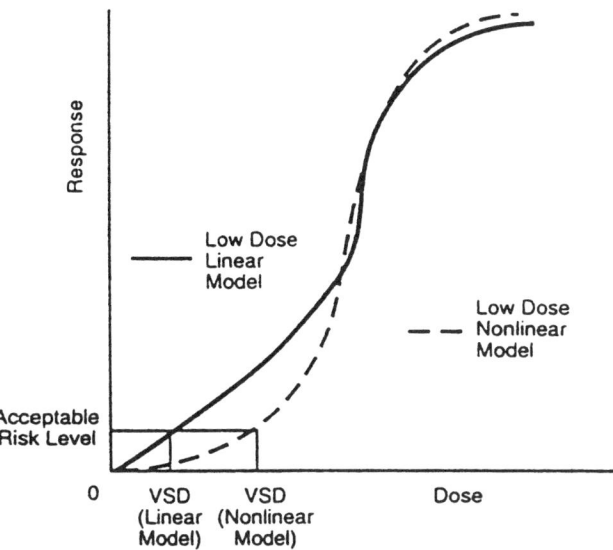

Fig. 4-3. The calculation of the virtually safe dose (VSD) with a low-dose linear and a low-dose nonlinear mathematical model.

Acceptable Daily Intake

The above discussion deals primarily with carcinogenicity. Some of the colorants have been shown to be carcinogenic, but in nearly every case the colorants suspected of being carcinogenic have been banned from use in food. Consequently, the most useful index of safety is the ADI (Fig. 4-2). The ADI is usually expressed as milligrams of the test substance per kilogram of body weight per day. Expressing ADI on the basis of body weight minimizes the size differences between animals. Expressing the dose as milligrams per day minimizes the differences in life expectancy between animals. Also, the ADI considers contributions from all aspects of the diet. It is used in the United States by the FDA and worldwide by the World Health Organization (WHO). The U.S. Environmental Protection Agency (EPA) prefers the term *reference toxicity dose (rTD)*, which does not imply acceptability. In most cases the ADI and the rTD are identical. The EPA also chose to replace the 100-fold safety factor with an "uncertainty factor."

The ADI values are usually determined from chronic toxicity studies, not acute toxicity experiments. The chief parameter in acute studies is the LD_{50}, which is defined as the dosage that results in the death of 50% of the experimental animals. LD_{50} values can be calculated for almost anything. For example, the LD_{50} value for starch with rats is

Reference toxicity dose (rTD)—Term used by the U.S. Environmental Protection Agency rather than ADI.

about 168 g/kg, at which point gastric rupture occurs, followed by death within a few hours (3). The parameter may be modified to include a time factor; for instance, LD_{50} (100 days) is the dosage that results in the death of 50% of the animals within 100 days. The LD_{50} (100 days) for sucrose for male albino rats is between 80 and 95 g/kg per day; death usually occurs as a result of osmotic imbalances. The LD_{50} is a useful value for judging acute toxicity, but it is not much help in long-term, chronic toxicity studies. It is not currently used as much as it was in the past.

In the United States, the ADI or rTD is converted to *daily intake*, which is generally based on a hypothetical average person who weighs 70 kg (4). For example, synthetic β-carotene has an ADI of 5 mg/kg per day. Therefore, the daily intake for β-carotene for a human would be 5 x 70 = 350 mg/day.

European Union

The principles that form the basis of food law are similar throughout the world, but the specifics vary considerably between countries. This is not surprising, considering that the expectations of countries differ and the interpretations, particularly of safety data, allow for considerable variation. The European Union (EU) is guided by three main principles: 1) protection of the health of the consumer, 2) prevention of fraud, and 3) removal of nontariff barriers to intracommunity trade.

The third point was the subject of much effort when the European Economic Community (EEC, forerunner of the EU) was created in 1957. A need for "harmonization" of the laws between the various countries was recognized, with the awareness that each country had its own national customs, tastes, and social customs. Harmonization was legislated under Article 100 of the 1957 Treaty of Rome, which established two types of EEC directives. The horizontal type, such as the Colors Directive, has provisions that apply across the board. Vertical directives apply to specific foods only. Directives are drawn up by the European Commission, which may take into account the recommendations of the EU Scientific Committee for Food (SCF), the Codex Alimentarius Commission, and the Joint FAO/WHO Expert Committee on Food Additives *(JECFA)*. Obviously, there is considerable international influence on the last two. Regardless of the source of the directive, it has standing only when approved by individual countries.

The EU Colors Directive lists colorants deemed to be suitable for use in food and specifies the limits of impurities. The SCF recommended that colorants for which an ADI could not be specified not be permitted in foods. It made an exception for colorants normally present in foods and for situations in which normal consumption would not exceed the amount of colorants normally present in foods. All other colorants require testing for safety before approval for use in

Daily intake—Term used by the FDA to estimate the intake of a given compound from all sources.

JECFA—The Joint Expert Committee on Food Additives of the Food and Agriculture Organization and World Health Organization.

food. This interpretation lent support to the *natural* vs. *synthetic* categorization of colorants. Since each country can modify the directive according to its own agenda, it is not surprising that there are some differences in the interpretation of the "natural" concept. Italy is probably the most restrictive of the EU countries. Norway does not allow any synthetic colorants and allows natural colorants only in amounts deemed to be technologically necessary, even though there are no specified limits. Each colorant is identified by a specific *E number*. For example, anthocyanins are listed as E 163. Parker (5) published a list of colorants and their status in Europe and the United Kingdom.

United States

FOOD LAWS

Food laws in the United States are similar in concept to those in Europe and are governed by the food and drug acts. The first one of importance, the Food and Drug Act of 1906, has been modified several times. U.S. Food Inspection Decision 77, issued in 1907, established the principle of certification of color additives. A representative sample of each color is submitted to the Bureau of Chemistry of the USDA for conformance to the established purity specifications. The fee for certification is paid by the manufacturer whether the batch passes or not. The 1938 modification was one of the more important developments; its principles are still in effect. The 1938 Food, Drug and Cosmetic Act established the FD&C colorants (the initials stand for "food, drug and cosmetic"). There is also a *D&C* classification for colorants allowed in drugs and cosmetics. The act specified a list of colorants allowed in foods and presumably held to a higher standard of purity than those available for other industrial uses. Colorants used in food would have to be identified by their FD&C name, but colorants used for other industrial purposes could still be referred to by their common names. Certification became mandatory.

To further classify colorants, the Society of Dyers and Colorists in England and the American Association of Textile Chemists and Colorists in the United States developed a color index (CI) system. Each colorant was given a *CI number* and CI name. Colorants are also identified by their Chemical Abstract Service (*CAS*) Registry code number. The FDA Center for Food Safety and Applied Nutrition (CFSAN) uses a CAS-like code for those substances without a number assigned by CAS.

The 1960 Color Additive Amendment to the 1938 act specified two groups of colorants: certified color additives (*certified colorants*) and color additives exempt from certification (*exempt colorants*). The first group currently contains seven synthetic FD&C colorants and two synthetic colorants for restricted use, all of which are required to be certified to comply with the purity specifications of the FDA. The second group currently contains 26 colorants (Table 4-1) that are *not* re-

Natural colorant—A regulatory term used in other parts of the world, but not in the United States, to denote a colorant that exists in nature. The exact definition varies by country.

Synthetic colorant—A colorant that does not occur in nature.

E number—The number assigned to a colorant in Europe.

D&C colors—Colors permitted by FDA for use in drugs and cosmetics.

CI number—The color index number assigned by the American Association of Textile Chemists and Colorists.

CAS code number—A number assigned to a colorant by the Chemical Abstract Service Registry.

Certified colorants—Colorants required by law to be certified by the FDA, e.g., FD&C Red No. 40.

Exempt colorants—A group of 26 colorants not required to be certified.

TABLE 4-1. Regulatory Status of Colorants Exempt from Certification in the United States[a]

Color Additive	U.S. Food Use Limit	EU[b] Status	JECFA ADI[c] (mg/kg of body weight)
Algal meal, dried	GMP[d] for chicken feed	NL[e]	NE[f]
Annatto extract	GMP[g]	E 160b	0–0.65[g]
Dehydrated beets	GMP	E 162	NE
Ultramarine Blue	Salt for animal feed up to 0.5% by weight	NL	None
Canthaxanthin	Not to exceed 30 mg/lb of solid/semisolid food or pint of liquid food or 4.41 mg/kg of chicken feed	E 161g	None
Caramel	GMP	E 150	0–200
β-Apo-8′-carotenal	Not to exceed 15 mg/lb of solid or semisolid food or 15 mg/pt of liquid food	E 160a	0–5
β-Carotene	GMP	E160a	0–5
Carrot oil	GMP	NL	NE
Cochineal extract or carmine	GMP	E 120	0-5
Corn endosperm oil	GMP for chicken feed	NL	NE
Cottonseed flour, toasted partially defatted cooked	GMP	NL	NE
Ferrous gluconate	GMP for ripe olives only	NL	NE
Fruit juice	GMP	NL	NE
Grape color extract	GMP for nonbeverage foods	E 163	NE
Grape skin extract (Enocianina)	GMP for beverages	E 163	0–2.5
Iron oxide, synthetic	Pet food up to 0.25%	E 172	0–2.5
Paprika	GMP	E 160c	None allocated
Paprika oleoresin	GMP	E 160c	Self-limiting as spice
Riboflavin	GMP	E 101	0–0.5
Saffron	GMP	NL	Food ingredient
Tagetes (Aztec marigold) meal and extract	GMP for chicken feed	NL	NE
Titanium dioxide	Not to exceed 1% by weight of food	E 171	Not specified
Turmeric	GMP	E 100 (curcumin)	Temporary ADI not extended
Turmeric oleorisin	GMP	E 100 (curcumin)	Temporary ADI not extended
Vegetable juice	GMP	NL	NE

[a] Adapted from (6).
[b] European Union.
[c] The acceptable daily intake (ADI), according to the Joint Expert Committee on Food Additives (JECFA) of the Food and Agriculture Organization and World Health Organization.
[d] According to good manufacturing practices.
[e] Not listed.
[f] Not evaluated.
[g] Calculated as bixin.

quired by FDA to be analyzed and certified as each batch was made. The amendment also established the principle of provisional vs. permanent listing of colorants. This was to allow for completion of new toxicity texts.

Many of the colorants that would be called "natural" in other countries are on the list of 26. FDA does not accept the concept of "natural" vs. "synthetic" and requires that compounds in both groups be subjected to the same standards of safety. Actually, in practice, the safety studies required for some of the 26 exempt colorants were considerably less than those required for the seven certified ones. This approach is understandable if one accepts the concept of the Decision Tree approach for determining safety, as suggested by the Food Safety Council (2). See Box 4-1.

> **Box 4-1. The Decision Tree**
>
> The Decision Tree, developed by a committee of the Food Safety Council, is a way of analyzing the risks associated with a particular food by asking a series of questions about the risks. One answer to each question causes the food to be rejected; another causes it to be accepted and subjected to the next question. Thus, the process branches like a tree.
>
> The first question concerns the risk of the food causing acute toxicity. If the risk is so high as to be socially unacceptable, the food is rejected. If the risk is socially acceptable, the food is provisionally accepted and another type of risk is considered. At each of three more branch points, the acceptability of the risk is questioned, and the food is rejected, accepted, or passed along. The final branch point leads only to acceptance or rejection. This process prioritizes the risks and leads to a clear decision about the acceptability of the food's risks and therefore the acceptability of the food itself.

The Nutritional Labeling and Education Act of 1990 mandated that certified color additives be specifically declared by their individual names, but the requirements for exempt colorants were left unchanged. Exempt colorants can still be declared generically as "artificial color" or any other specific or generic name for the colorant. However, the term "natural" is prohibited because it may lead the consumer to believe that the color is derived from the food itself. There is no such thing as a "natural" colorant in the United States. Detailed accounts of the safety of the noncertified colorants have recently been published (6,7).

DELANEY CLAUSE

The United States is unique in the world in that it has had to deal with a restriction, the *Delaney Clause* (Box 4-2), that was passed by

Delaney Clause—A clause in the 1958 Food Additive Amendment forbidding the use of a substance if, after appropriate tests, any part of it was shown to cause cancer in humans or animals.

Box 4-2. Highlights of Regulations for Color Additives in the United States

Resources. The use of all food colors in the United States is governed by the *Code of Federal Regulations* (CFR), which is a codification of the rules issued by agencies of the federal government. The CFR is divided into 50 titles; Title 21 is assigned to the Food and Drug Administration (FDA). Within Title 21 (abbreviated 21 CFR), Parts 70–82 deal with color additives and include 1) general provisions, including definitions and labeling; 2) a listing of all FDA-approved colors, including both colors that are certified (FD&C) and those that are exempt from certification; 3) purity specifications and requirements for each color; and 4) uses of and restrictions for each color. If a color additive is not specifically included somewhere in these sections, it may not be added to food, drug, or cosmetic products or medical devices that will be used in the United States. CFR Parts 70–82 are recommended reading for everyone who deals with color additives because they spell out the rules under which both industry and FDA operate.

One can monitor changes in the regulations by reading the *Federal Register*. New regulations that are being contemplated must be published there to allow time for comment, and a new regulation does not become official until it has been published in the *Federal Register*. However, following regulatory changes in this way is often not practical for people who use color additives, and questions regarding colors and their regulatory status should be referred to companies with strong technical service and regulatory departments.

The International Association of Color Manufacturers (IACM) is a trade association devoted to protecting the use of safe colorants worldwide through strong scientific and legal analysis and representation. Many color manufacturers support the IACM financially and with contributions of labor.

Certified Colors. These are synthetic colors requiring certification by the FDA for compliance with established purity specifications. They are divided into three groups: FD&C, D&C, and External D&C colors. A fee for certification is paid to the FDA for each batch of color submitted, whether it passes or not. These colors come in basically two forms—dyes and lakes. Most of the dyes are currently permanently listed, while most of the lakes are provisionally listed.

Exempt Colors. Natural colors are referred to by FDA as "exempt from certification" and are covered in 21 CFR Part 73, which includes specifications for high levels of purity. However, unlike the requirements for certified colors, a sample of each batch of exempt colorant does not need to be sent to FDA for certification. FDA does not consider any color to be natural if it is added to a product in which it is not normally present. For example, strawberry juice used to color strawberry ice cream is considered "natural," whereas beet juice used to color strawberry ice cream is not natural.

Although 26 exempt colors are permitted for food use under 21 CFR Part 73, careful consideration must be given to color choice so that only permitted and defined colors are used and then only for their permitted applications. For example, although Grape Color Extract is listed in 21 CFR Subpart A—Foods, it can be used only for coloring nonbeverage foods. Grape Skin Extract, which is also listed in this section, is restricted to use in coloring beverages only. Another example listed in this section is Tagetes Meal, which is permitted as colorant for chicken feed only.

Labeling. In the United States, since 1993, certified color additives, including lakes, must be listed in a food product's ingredient statement by name. It is not necessary to include the "FD&C" prefix or the term "No.," although the term *lake* must be used (e.g., Blue 1 Lake). Although the industry refers to exempt colors as "natural" colors, use of this term is prohibited on an ingredient statement. According to 21 CFR section 101.22, several labeling options for exempt colors are permitted: "color added," "artificial color added," "colored with . . . " (e.g., colored with annatto), " . . . color" (e.g., beet juice color), or " . . . (color)" (e.g., cochineal extract (color)).

Congress in the 1958 Food Additives Amendment, one of three amendments to the 1938 Food, Drug and Cosmetics Act. Briefly, the clause states that no additive shall be added to food

> if it is found, after tests which are appropriate for the evaluation of safety of food additives, to induce cancer in man or animals.

Unfortunately, Congressman Delaney did not anticipate the ingenuity of modern chemists in pushing back the limits of sensitivity. Today, analyses are available in which levels as low as one molecule can be detected. The ability to detect infinitely small amounts has far outstripped the ability of toxicologists to interpret the results. The end result has been an unworkable regulatory clause. Scientific opinion in the United States is almost unanimous in declaring that the Delaney Clause is hopelessly obsolete and should be declared invalid.

Current legislative debate concerns modifying or eliminating the clause and replacing it with some form of *de minimis*, a legal concept that comes from the term *de minimis curat lex*. Freely translated, this term means that the law does not concern itself with trifles. The Food Quality Protection Act, signed into law in August 1996, eliminated the Delaney Clause from the pesticide area but did empower EPA to set a new standard for pesticides that may possibly be stricter than the old tolerances. Under the new law, EPA must determine that the new tolerance is safe, with "safe" being defined as reasonable certainty of causing no harm. However, the Delaney Clause still applies to colorants.

SOURCES OF INFORMATION

Testing for safety of food additives in the United States is described in detail in an FDA publication entitled *Toxicological Principles for the Safety Assessment of Direct Food and Color Additives Used in Food*. The 1993 draft is commonly known as "Redbook-2."

FDA has also made it easy to determine the status of a food additive by publishing a book entitled *Everything Added to Food in the United States* (8). The information in the book was derived from FDA files and describes a total of 2,922 food additives, of which 1,755 are regulated food additives, including direct, secondary direct, color, and Generally Recognized as Safe (*GRAS*) additives. The book also contains administrative and chemical information on 1,167 additional substances. It is arranged alphabetically, and each compound is identified by its CAS number. The appropriate section in Title 21 of the *U. S. Code of Federal Regulations* (*CFR*) for each substance is also listed. For example, paragraph (e) of CFR 21-178.3297 provides a description and a long list of compounds that

> may be safely used as colorants in the manufacture of articles or components of articles intended for use in producing, manufacturing, packing, processing, preparing, treating, packaging, transporting or holding food, subject to the provisions and definitions set forth in this section . . . . The term colorant means a dye, pig-

De minimis—A concept used by regulatory agencies to describe a substance with a negligible risk.

GRAS—A Food and Drug Administration regulatory status meaning Generally Recognized as Safe.

CFR—The *Code of Federal Regulations*, the compilation of all U.S. laws.

ment, or other substance that is used to impart color or to alter the color of a food-contact material, but that does not migrate into the food in amounts that will contribute to that food any color apparent to the naked eye. For the purposes of this section, the term "colorant" includes substances such as optical brighteners and fluorescent whiteners, which may not themselves be colored, but whose use is intended to affect the color of a food-contact material.

The above section is very general, but the sections on individual colorants are more specific. For example, talc has only one entry (21 CFR 182.90) and FD&C Red No. 40 has three (21 CFR 74.340, 74.1340, and 74.2340).

The data in *Everything Added to Food in the United States* is also listed in a much broader publication entitled *Food Additives: Toxicology, Regulation, and Properties* (9), which is available as a CD ROM.

The emphasis on food safety possibly has had another influence, namely the worldwide trend toward the use of colorants derived from natural sources. A survey of food colorant patents (10) in the 15-year period from 1969 to 1984 found 356 patents on colorants from natural sources and 71 on synthetics. The ratio is even more lopsided when one considers that 21 of the 71 were on colorants linked to a polymer background (the Dynapol concept), an approach that is no longer being considered. Little effort worldwide is being devoted to the development of novel synthetic colorants, but considerable effort is being devoted to the development of colorants from natural sources.

References

1. Francis, F. J. 1986. Testing for toxicity. J. Sci. Food Agric. 4:10-13.
2. Food Safety Council. 1982. A proposed food safety evaluation process. Final report of the Board of Trustees. Nutrition Foundation, Washington, DC.
3. Boyd, E. M. 1973. *Toxicity of Pure Foods*. CRC Press, Boca Raton, FL.
4. Rodricks, J. V. 1992. *Calculated Risks: The Toxicity and Human Health Risks of Chemicals in our Environment*. Cambridge University Press, Cambridge, UK.
5. Parker, L. E. 1984. Regulatory approaches to food coloration. Chap. 1 in: Developments in Food Colours-2. Applied Science Publishers, London.
6. Hallagan, J. B., Allen, D. C., and Borzelleca, J. F. 1995. The safety and regulatory status of food, drug and cosmetic color additives exempt from certification. Food Chem. Toxicol. 33:515-528.
7. Francis, F. J. 1996. Safety of natural food colorants. In: *Natural Food Colorants*. G. A. F. Hendry and J. D. Houghton, Eds. Blackie Publishers, Glasgow, Scotland.
8. FDA. 1993. *Everything Added to Food in the United States*. CRC Press, Boca Raton, FL.
9. Clydesdale, F. M. 1997. *Food Additives: Toxicology, Regulation, and Properties*. CD ROM. CRC Press, Boca Raton, FL.
10. Francis, F. J. 1986. *Handbook of Food Colorant Patents*. Food and Nutrition Press, Westport, CT.

CHAPTER 5

FD&C Colorants

In This Chapter:
Safety of Colorants
Stability in Foods
Lakes

The history of synthetic colorants dates back to the discovery of the first synthetic dye, mauve, by Sir William Henry Perkin in 1856. Since then, over 700 colorants have been available to the paint, plastic, textile, and food industries. In 1907, about 80 colorants were offered for use in foods (1), and obviously very few had been tested for safety. At that time, Dr. Bernard Hesse, a German dye expert employed by the USDA, was asked to study the colorants available for use in foods. He concluded that only 16 of the 80 colorants were probably more or less harmless and recommended only seven for general use in food.

Most of Hesse's information was embodied in the Food and Drug Act of 1906. This act, together with Food Inspection Decision No. 76 in July 1907, put an end to the indiscriminate use of colorants in food. The new legislation allowed only colorants that were of known chemical structure and that had been tested for safety. The act also set up a system for certification of synthetic organic food colorants designed for use in foods. The certification of each batch included proof of identity and documentation of the levels of impurities. During the next three decades, 10 more colorants were added to the list.

The Federal Food, Drug and Cosmetic Act of 1938 forbade the use of any uncertified coal-tar color in any food, drug, or cosmetic shipped in interstate commerce, and it created three categories of coal-tar colors:

- FD&C colors: those certifiable for use in coloring foods, drugs, and cosmetics.
- D&C colors: dyes and pigments considered safe in drugs and cosmetics when in contact with mucous membranes or when ingested.
- External (Ext.) D&C colors: colorants not certifiable, because of their oral toxicity, for use in products intended for ingestion but considered safe for use in products applied externally.

Each colorant had to be identified by its specific FD&C, D&C, or Ext. D&C name. This act also stated that

> The Secretary [of Health, Education and Welfare] shall promulgate regulations providing for the listing of coal-tar colors which are harmless and suitable for food.

Safety of Colorants

The passage of the Federal Food, Drug and Cosmetic Act of 1938 drew public attention to the question of safety of colorants and led to the publication of a regulatory announcement (2) in 1940. This document listed specific colorants that could be used, with purity specifications and regulations pertaining to manufacture and sale.

In 1950, a dilemma developed, precipitated by two events. First, a number of children became sick after reportedly eating popcorn and candy colored with excessive amounts of colorant. Second, the FDA launched a new round of toxicity tests with animals fed much higher levels of colorants and for a longer time than any tests previously conducted. The results were unfavorable for FD&C Orange No. 1, FD&C Orange No. 2, and FD&C Red No. 32, and these were taken off the list of allowable colorants.

The FDA based its conclusions on the fact that the 1938 Act required coal-tar colors on the list to be "harmless and suitable for food." FDA interpreted "harmless" to mean harmless at any level. Food manufacturers argued that this interpretation was too strict and the FDA should set safe limits. The U.S. Supreme Court held that the FDA did not have the authority to set limits, and several more colorants were delisted, including FD&C Yellows Nos. 1–4.

It was obvious that the law was unworkable so, through the efforts of the certified color industry and FDA, a new law was passed. The Color Additive Amendment of 1960 allowed for provisional use of colorants pending completion of testing and, equally important, authorized the Secretary of Health, Education and Welfare to establish limits of use, thereby eliminating the "harmless per se" interpretation. Another feature allowed the Secretary to determine which colorants had to be certified and which would be exempt from certification. This was also the act into which the "Delaney Clause" (see Chapter 4) was inserted.

In view of the expense and time required for adequate testing, some of the colorants had "provisional" listing for a number of years, but this category for straight FD&C colorants eventually expired. Because of the expense involved, only commercial colorants of significant economic value were tested, so some colorants were delisted by default. Actually, this should not be a hardship because the nine currently allowed colorants (Table 5-1) should be sufficient in view of the opportunities for blending. Interestingly, of the seven colorants that Dr. Hesse recommended for general use in food, only two (FD&C Red No. 3 and FD&C Blue No. 2) remain on the current list of approved colorants. FD&C Blue No. 2 was listed provisionally until it was given final approval in 1987. Similarly, FD&C Yellow No. 6 was finally approved in 1986. Formulas for the nine colorants are shown in Figure 5-1.

The FDA has been petitioned to add a synthetic colorant, D&C Red No. 28 (Phloxene B, CI 45410), to the approved list and is currently beginning the review process. There is some interest in approving

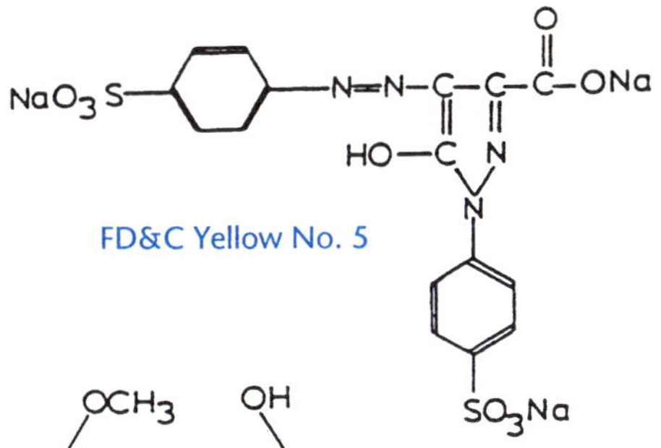

FD&C Yellow No. 5

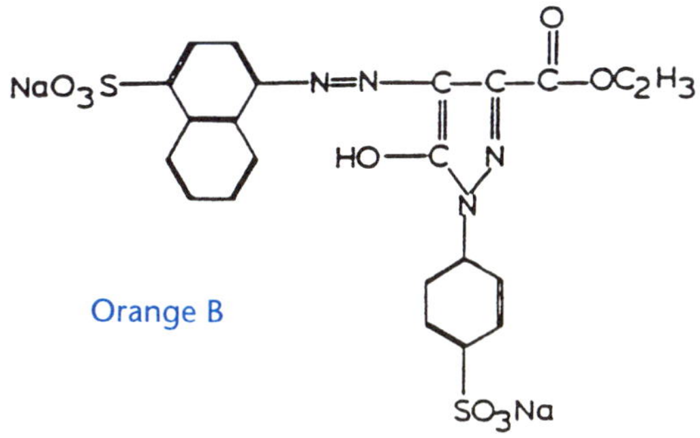

Orange B

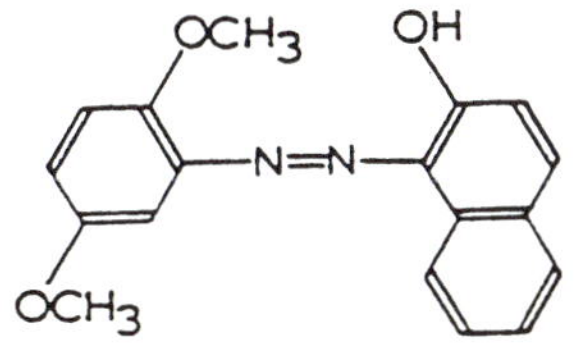

Citrus Red No. 2

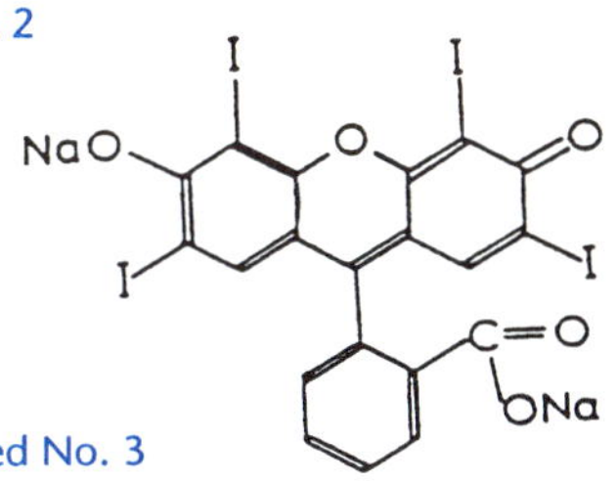

FD&C Red No. 3

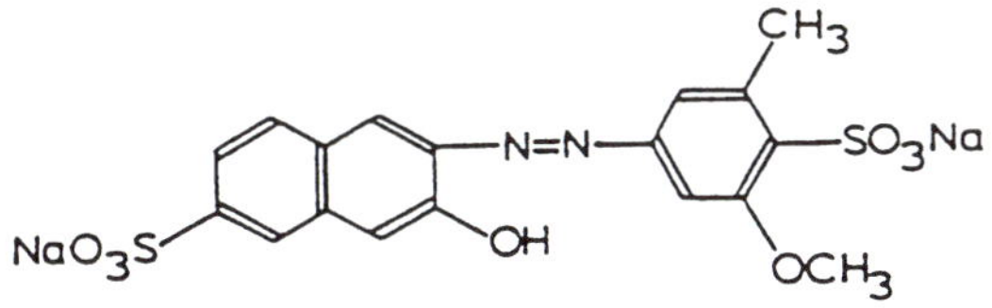

FD&C Red No. 40

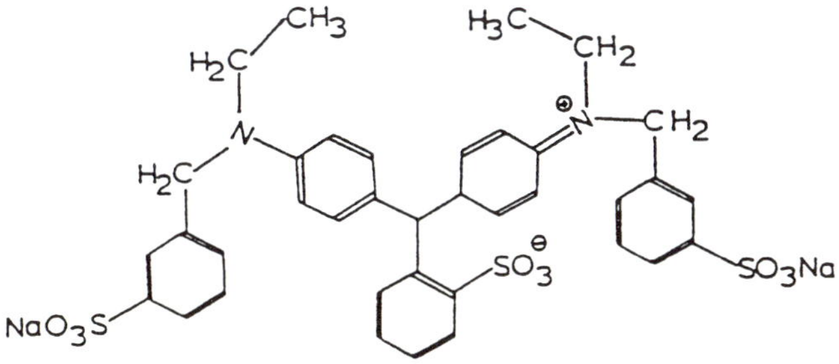

FD&C Blue No. 1

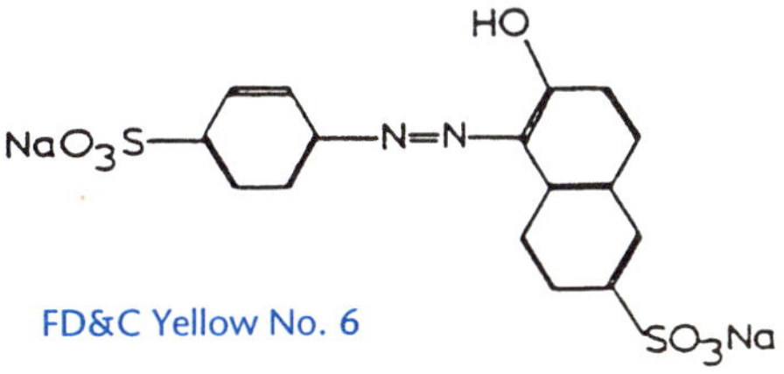

FD&C Yellow No. 6

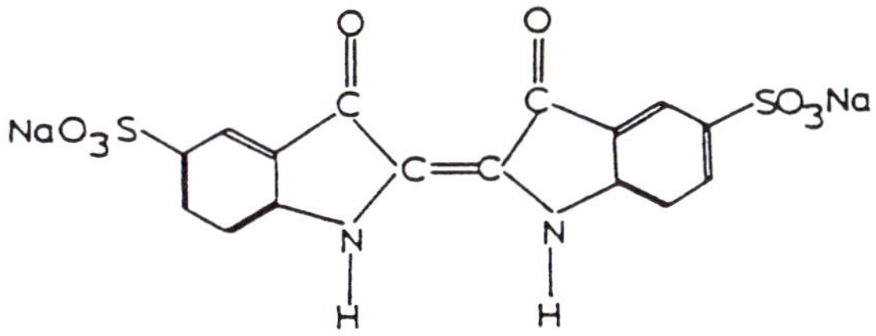

FD&C Blue No. 2

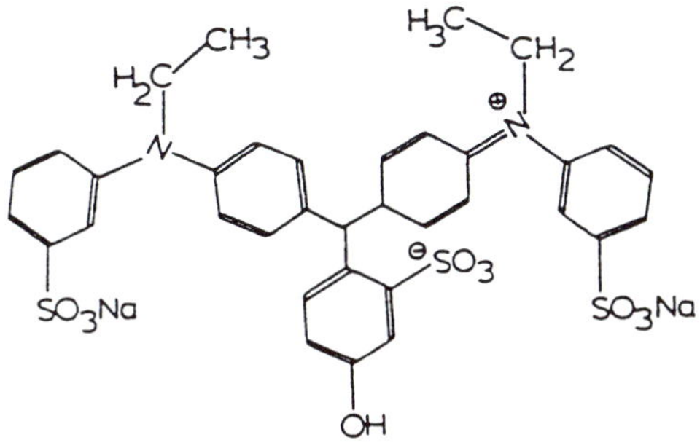

FD&C Green No. 3

Fig. 5-1. Structures of the nine certified colorants.

two other synthetic colorants, D&C Yellow No. 10 and carmoisine for food use. In the United States, D&C Yellow No. 10 (Quinoline Yellow, CI 47005) is predominantly the monosulfomonosodium salt of disulfonic acid, as opposed to the Quinoline Yellow of Europe (EU No. E104), which is primarily the disodium salt of disulfonic acid. If Quinoline Yellow is ever approved for food use in the United States, it probably will be named D&C Yellow No. 12. Carmoisine (azorubine, CI 14720) is composed of napthionic acid and several derivatives and has been the subject of several toxicological studies, particularly in the United Kingdom. None of these three colorants is approved for food use in the United States at this time.

Stability in Foods

The nine colorants listed in Table 5-1 are generally sufficient to enable a formulator to create any desired color by mixing colorants, as discussed in Chapter 3. However, each colorant may behave differently in each food system; therefore, prediction from theoretical knowledge of the colorant is difficult. Blending must be done under actual food conditions.

Table 5-2 shows the stability of FD&C colorants under a variety of conditions. Tables 5-3 and 5-4 show stability under acid conditions, and Table 5-5 shows changes under basic conditions. With sugars, the FD&C colorants are unchanged in the presence of 10% dextrose or sucrose except for FD&C Blue No 2, which shows considerable fading in 10% dextrose and slight fading in 10% sucrose.

The tables refer to changes in liquid media in the presence of a variety of chemicals, but solid media may show different changes, possibly because of absorption of the colorant on the surface of the media. For example, solid starch or sugar is often used as a chromatographic reagent, and the absorption of the colorant on the surface changes the color of the mixture. However, if color readings are

TABLE 5-1. Colorants Currently Approved for Use in the United States

FDA Name	Common Name	CI Number	Year Listed	Chemical Class	Hue
FD&C Blue No. 1	Brilliant Blue	42090	1929	Triphenylmethane	Greenish blue
FD&C Blue No. 2	Indigotine	73015	1907	Indigoid	Deep blue
FD&C Green No. 3	Fast Green	42053	1927	Triphenylmethane	Bluish green
FD&C Yellow No. 5	Tartrazine	19140	1916	Pyrazolone	Lemon yellow
FD&C Yellow No. 6	Sunset Yellow	15985	1929	Monoazo	Reddish yellow
FD&C Red No. 3	Erythrosine	43430	1907	Xanthine	Bluish pink
FD&C Red No. 40	Allura Red	16035	1971	Monoazo	Yellowish red
Orange B[a]	Orange B	…	1966	Monoazo	
Citrus Red No. 2[b]	Citrus Red No. 2	…	1959	Monoazo	

[a] Allowed only on the surfaces of sausages and frankfurters at concentrations up to 150 ppm by weight.
[b] Allowed only on the skins of oranges, not intended for processing, at concentrations up to 2 ppm by weight of the whole fruit.

TABLE 5-2. Stability[a] of FD&C Colorants Under Various Conditions[b]

Colorant Name	Condition					
	Light	Acids	Alkalies	Sulfur Dioxide	Ascorbic Acid	Heat
FD&C Blue No. 1	4	5	4	4	4	5
FD&C Blue No. 2	1	2	2	2	3	4
FD&C Green No. 3	4	5	2	4	4	4
FD&C Yellow No. 5	5	5	5	4	2	4
FD&C Yellow No. 6	4	5	5	4	2	3
FD&C Red No. 3	2	2	5	5	2	5
FD&C Red No. 40	5	5	4	2	2	3

[a] Stability: 1 = very poor, 5 = good.
[b] Modified from (3).

TABLE 5-3. Stability of FD&C Colorants at Various pH Values[a]

Colorant Name	pH			
	3	5	7	8
FD&C Blue No. 1	sf	vsf	vsf	vsf
FD&C Blue No. 2	af	af	cf	fc
FD&C Green No. 3	sf	vsf	vsf	sf (bluer)
FD&C Yellow No. 5	naf	naf	naf	naf
FD&C Yellow No. 6	naf	naf	naf	naf
FD&C Red No. 3	ins	ins	naf	naf
FD&C Red No. 40	naf	naf	naf	naf

[a] sf = slight fade after one week, vsf = very slight fade after one week, af = appreciable fade after one week, cf = considerable fade after one week, fc = completely faded, naf = no appreciable fading, ins = insoluble.

TABLE 5-4. Stability of FD&C Colorants in the Presence of Acids[a]

Colorant Name	10% Citric Acid	10% Acetic Acid	10% Malic Acid	10% Tartaric Acid
FD&C Blue No. 1	naf	naf	naf	naf
FD&C Blue No. 2	fc	fc	cf	cf
FD&C Green No. 3	naf	naf	sf	sf
FD&C Yellow No. 5	naf	naf	naf	naf
FD&C Yellow No. 6	naf	naf	naf	naf
FD&C Red No. 3	ins	ins	ins	ins
FD&C Red No. 40	naf	naf	naf	naf

[a] sf = slight fade after one week, cf = considerable fade after one week, fc = completely faded, naf = no appreciable fading, ins = insoluble.

TABLE 5-5. Stability of FD&C Colorants in the Presence of Alkalies[a]

Colorant Name	10% Sodium Bicarbonate	10% Sodium Carbonate	10% Ammonium Hydroxide	10% Sodium Hydroxide
FD&C Blue No. 1	sf	fc	cf	fc
FD&C Blue No. 2	fc	fc	fc	Yellower
FD&C Green No. 3	naf	cf (bluer)	cf (bluer)	fc
FD&C Yellow No. 5	naf	naf	naf	cf
FD&C Yellow No. 6	naf	naf	naf	sf
FD&C Red No. 3	naf	sf	sf	fc
FD&C Red No. 40	Slightly bluer	Bluer	Bluer	Much bluer

[a] sf = slight fade after one week, cf = considerable fade after one week, fc = completely faded, naf = no appreciable fading.

TABLE 5-6. Solubility of FD&C Colorants in Water (g/100 ml)

Colorant Name	Temperature, °C		
	2	25	60
FD&C Blue No. 1	20	20	20
FD&C Blue No. 2	0.8	1.6	2.2
FD&C Green No. 3	20	20	20
FD&C Yellow No. 5	3.8	20	20
FD&C Yellow No. 6	19	19	20
FD&C Red No. 3	9	9	17
FD&C Red No. 40	18	22	26

TABLE 5-7. Solubility of FD&C Colorants in Ethanol (g/100 ml)

Colorant Name	Alcohol Concentration and Temperature							
	25%		50%		75%		100%	
	25°C	60°C	25°C	60°C	25°C	60°C	25°C	60°C
FD&C Blue No. 1	20	20	20	20	20	20	0.15	0.15
FD&C Blue No. 2	0.5	0.6	0.3	0.4	0.07	0.07	0	0.008
FD&C Green No. 3	20	20	20	20	10	20	0.01	0.01
FD&C Yellow No. 5	12	17	4	8	1	1	0	0
FD&C Yellow No. 6	10	15	3	4	0.3	0.3	0	0
FD&C Red No. 3	8	8	1	1	0.6	0.8	0	0
FD&C Red No. 40	9	22	1.3	5	0.3	0.9	0	0.05

taken on the same media with different colorants or different concentrations of the same colorant, the laws of color blending as discussed in Chapter 3 should still apply. This only serves to reinforce the fact that the color readings should be taken on the final product.

Tables 5-6 through 5-9 show solubility in water, ethanol, propylene glycol, and glycerol, respectively. Table 5-10 shows the blends of five colorants used to create a number of popular colors. Obviously, many other combinations are possible.

TABLE 5-8. Solubility of FD&C Colorants in Propylene Glycol (g/100 ml)

Colorant Name	Glycol Concentration and Temperature							
	25%		50%		75%		100%	
	25°C	60°C	25°C	60°C	25°C	60°C	25°C	60°C
FD&C Blue No. 1	20	20	20	20	20	20	20	20
FD&C Blue No. 2	0.6	2	0.4	0.4	0.4	0.4	0.1	0.1
FD&C Green No. 3	20	20	20	20	20	20	20	20
FD&C Yellow No. 5	20	20	12	20	10	13	7	7
FD&C Yellow No. 6	20	20	7	13	2	3	2	2
FD&C Red No. 3	7	9	7	9	16	16	20	20
FD&C Red No. 40	18	22	7	10	2	3	1.5	1.7

TABLE 5-9. Solubility of FD&C Colorants in Glycerol (g/100 ml)

Colorant Name	Glycerol Concentration and Temperature							
	25%		50%		75%		100%	
	25°C	60°C	25°C	60°C	25°C	60°C	25°C	60°C
FD&C Blue No. 1	20	20	20	20	20	20	20	20
FD&C Blue No. 2	1.0	1.5	1.0	1.5	1.0	1.0	1.0	1.0
FD&C Green No. 3	20	20	20	20	20	20	20	20
FD&C Yellow No. 5	20	20	20	20	20	20	18	18
FD&C Yellow No. 6	20	20	20	20	20	20	20	20
FD&C Red No. 3	14	19	16	16	20	20	20	20
FD&C Red No. 40	20	20	12	14	5	9	3	8

TABLE 5-10. Blends of Colorants (parts by weight) to Produce Specific Colors

Color	FD&C Colorant				
	Blue No. 1	Red No. 3	Red No. 40	Yellow No. 5	Yellow No. 6
Strawberry		5	95		
Raspberry	5	75			20
Lime green	3			97	
Mint green	25			75	
Orange			25	20	55
Peach		21		64	
Grape	20		80		
Black cherry	5		95		
Butterscotch	3		22	57	18
Caramel	6	21		64	9
Cola	5		25		70
Chocolate	10		45	45	
Licorice (black)	36		22		42

Lake—A water-soluble colorant precipitated on a base of aluminum or calcium salts.

Lakes

Lakes are colorants prepared by precipitating a soluble colorant onto an insoluble base or substratum. A variety of bases such as alumina, titanium dioxide, zinc oxide, talc, calcium carbonate, and aluminum benzoate are approved for D&C colorants, but only alumina is permitted as the substrate for manufacturing FD&C lakes.

The process of manufacture is fairly simple. First, a substratum is prepared by adding sodium carbonate to a solution of aluminum sulfate. Next a certified colorant is added to the slurry, followed by aluminum sulfate to convert the colorant to an aluminum salt, which is absorbed onto the surface of the alumina. The slurry is then washed, dried, and ground to a fine powder.

The colored powder can be marketed as is or mixed with a diluent such as hydrogenated vegetable oil, coconut oil, propylene glycol, glycerol, sugar syrup, or other media appropriate for food consumption or for printing food wrappers. Diluents for both pure colorants or lakes must be either GRAS or on the list approved by the FDA *Code of Federal Regulations*. Table 5-11 lists four non-GRAS components approved for general use as diluents. There are also a number of diluents approved for specialty use. The *Code of Federal Regulations*, Chapter 13, lists 13 diluents used as inks for marking food supplements in tablet form, gum, and confectionery. It also lists 12 for use in marking fruits and vegetables and 13 for coloring shell eggs.

TABLE 5-11. Non-GRAS Diluents[a] in Color Additive Mixtures

Substance	Definition and Specifications	Restrictions
Castor oil	USP XVI specifications	Not more than 500 ppm in the finished food. Must bear adequate labeling.
Dioctylsodium sulfosuccinate	As set forth in Sec. 172.810 of 21 CFR 73.1. Subpart A–Foods	Not more than 9 ppm in the finished food. Labeling of color additive mixtures containing dioctylsodium sulfosuccinate shall bear adequate directions for use that will result in a food meeting this restriction.
Disodium ethylene-diaminetetraacetate (EDTA)	Contains disodium ethylenediaminetetraacetate dihydrate (CAS Reg. No. 6381-92-6) Food Chemicals Codex, 3rd ed., p. 104, 1981	May be used in aqueous solutions and aqueous dispersions as a preservative and sequestrant in color additive mixtures intended only for ingested use. The color additive mixture may contain not more than 1% by weight of the diluent.
Calcium disodium EDTA	Contains calcium disodium ethylenediaminetetraacetate dihydrate (CAS Reg. No. 6766-87-6) Food Chemicals Codex, 3rd ed., p. 50, 1981	Same as for disodium EDTA

[a] These substances are approved for general use. Diluents may also be substances on the GRAS list.

Lake colorants have many advantages, primarily because they are insoluble in most solvents, including water. They have high opacity, are easily incorporated into dry media, and have superior stability to light and heat. Lakes are effective colorants for candy and pill coatings since they do not require removal of water before processing. They are particularly effective for coloring hydrophobic foods such as fats and oils, confectionery products, bakery products, salad dressings, chocolate substitutes, and other products in which the presence of water is undesirable. Lakes made from some water-soluble FD&C colorants may show some bleeding when formulated in foods outside the pH range 4.5–8.0, but few foods are in this pH range. Lakes have been used to color food packaging materials, including lacquers, plastic films, and inks, from which straight water-soluble colorants would be soon leached. Some of the differences between lakes and pure colorants are shown in Table 5-12.

Lakes can be prepared with a range of pure colorant content, but 10–40% is usual. They impart their color by *dispersion* of solid particles in the food media; their colorimetric properties are highly dependent on the conditions employed in their manufacture, which include particle size, crystal structure, content of FD&C colorant, water content, etc.

TABLE 5-12. Differences in Properties Between Lakes and Pure Colorants

Property	Lakes	Pure Colorants
Solubility	Insoluble in most solvents	Soluble in water, alcohol, propylene glycol, and glycerol
Method of coloring	Dispersion	Solution
Pure colorant content	10–40%	90–93%
Rate of use	0.1–0.3%	0.01–0.03%
Particle size	Approximately 5 μm	74–1,200 μm
Stability		
Light	Better	Good
Heat	Better	Good
Coloring strength	Not proportional to pure colorant content	Directly proportional to pure colorant content
Hue	Varies with pure colorant content	Less variation with pure colorant content
Cost	More expensive	Less expensive

The FDA listed the lakes provisionally in 1959. The data available at that time supported provisional but not permanent listing. The FDA needed more information on definitions, nomenclature, safety, specifications, and analytical methods. Although more information is available today, all lakes, with the exception of FD&C Red No. 40, are still provisionally listed.

Dispersion—The suspension of very small particles in a liquid media. If the particles are small enough, the suspension is relatively stable.

References

1. Marmion, D. C. 1984. *Handbook of U.S. Colorants for Foods, Drugs, and Cosmetics*, 2nd ed. (3rd ed., 1991). John Wiley & Sons, New York.
2. FDA. 1940. Service and Regulatory Announcement, Food, Drugs and Cosmetics No. 3. U.S. Food and Drug Administration, Washington, DC.
3. Francis, F. J. 1985. Pigments and other colorants. In: *Food Chemistry*, 2nd ed. O. R. Fennema, Ed. Marcel Dekker, New York.

CHAPTER 6

Carotenoids

In This Chapter:
Annatto
Saffron
Paprika
Tagetes
Lycopene
Miscellaneous Carotenoid Extracts
Synthetic Carotenoids
Health Aspects

The carotenoids are probably the best known of the food colorants and certainly are one of the largest groups of pigments produced in nature. They are very widespread; over 100,000,000 tons are produced annually in nature. Most of this amount is in the form of fucoxanthin in algae in the ocean and the three main carotenoids of green leaves: lutein, violaxanthin, and neoxanthin. Other pigments predominate in certain plants, such as lycopene in tomatoes, capsanthin and capsorubin in red peppers, and bixin in annatto. Colorant preparations have been made from all of these, and obviously the composition of the colorant extracts reflects the profile of the starting material. Bauernfeind (1) described many of the colorant preparations in detail.

Carotenoids can occur in nature in four states: 1) as crystals or amorphous solids in solution or associated with lipid media, 2) as esters of fatty acids (e.g., the lauric acid esters of capsanthin in peppers), 3) in combinations with sugars (e.g., the gentiobioside of crocetin), and 4) in combination with protein, where the carotenoid stabilizes the molecule (e.g., the combination of astaxanthin with protein forms the greenish-blue pigment in the shells of lobsters).

β-Carotene can be used as an example of the structure of carotenoids. It has β-ionone structures at each end of the molecule joined by a conjugated double bond system. The structure and numbering system for β-carotene are shown in Figure 6-1. The modifica-

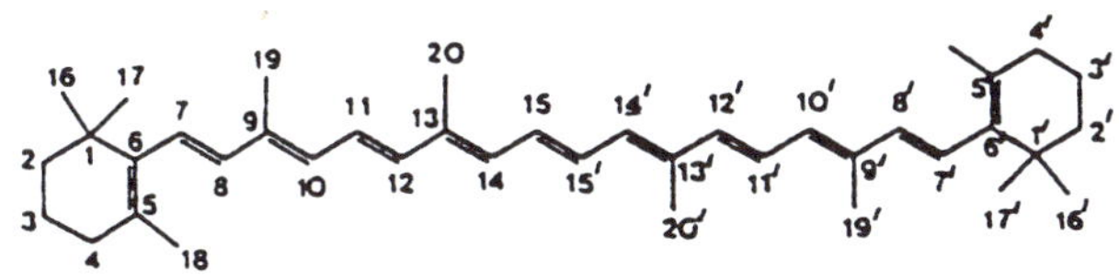

Beta-carotene

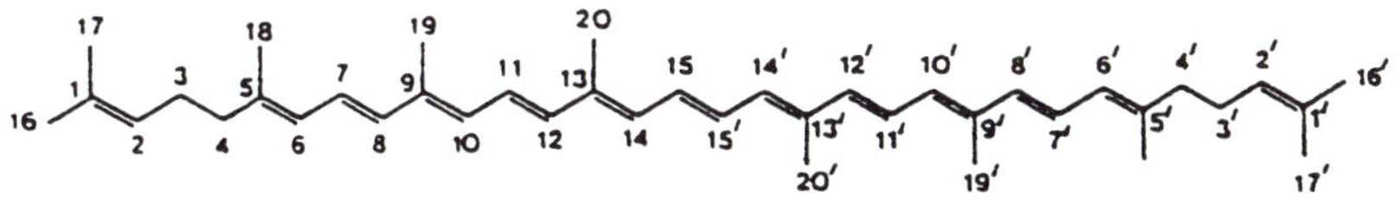

Lycopene

Fig. 6-1. Structure and numbering sequence for β-carotene and lycopene.

tions on each ionone structure vary greatly—over 600 carotenoids are known to occur in nature. The color of the carotenoid pigment is a function of the number of conjugated double bonds in the molecule. For example, phytofluene (Fig. 6-2) has only five conjugated bonds and is colorless. Zeta-carotene has seven and is yellow. Neurosporene has nine and is orange. Lycopene (Fig. 6-1) has 11 and is red. All four of these pigments occur in tomatoes. Extension of the conjugation system into the ring structure does influence the color but not to the same extent as the open structure of lycopene; thus β-carotene with 11 double bonds is orange. Canthaxanthin (Fig. 6-2) also has 11 conjugated bonds extending into the ring structure, but the ring keto group shifts the visual color to the red. The symmetry of the canthaxanthin molecule also contributes to its stability.

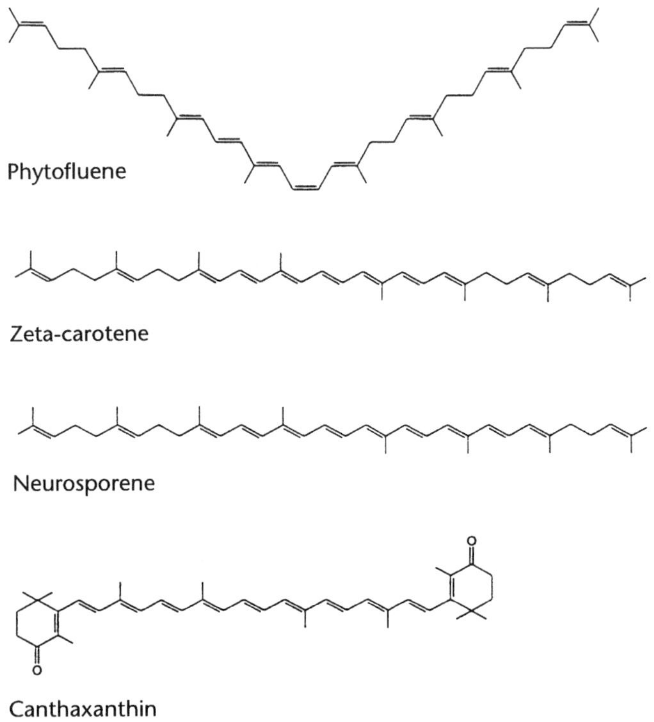

Fig. 6-2. Selected structures for the carotenoids in plants.

Annatto

Annatto (CI Natural Orange 4, CI No. 75120, EU No. E 160b) is found in the outer layers of the seeds of the shrub *Bixa orellana*, a tropical plant grown in South America, India, East Africa, the Philippines, and the Caribbean. Peru and Brazil are the dominant sources of supply. It is one of the oldest colorants, having been used in antiquity for coloring foods, cosmetics, and textiles. It has been used for over 100 years in the United States and Europe, primarily as a colorant for dairy products. The colorant is prepared by leaching the seeds, using gentle mechanical friction along with various solvents, including vegetable oil, fats, and aqueous alkali and alcoholic solutions. Depending on the application, the crude extract may be refined by precipitation with acids and/or recrystallization. Spray-dried powders are also available in both water-soluble and oil-soluble forms. The FDA-approved procedures for extraction are specified in 21 CFR section 73.30.

The pigments in annatto are a mixture of bixin and norbixin. Bixin, the monomethyl ester of a dicarboxylic carotenoid, is the naturally occurring form, and norbixin is the saponified form, a dicarboxylic derivative of the same carotenoid (Fig. 6-3). Both bixin and norbixin normally occur in the *cis* form, but small amounts of the more stable *trans* form are formed on heating. A yellow degradation product, termed C_{17} yellow pigment, is also produced on heating. The *cis* forms are redder than the *trans* forms or the C_{17} compound; thus, a source of red and yellow pigments is available. The carboxylic

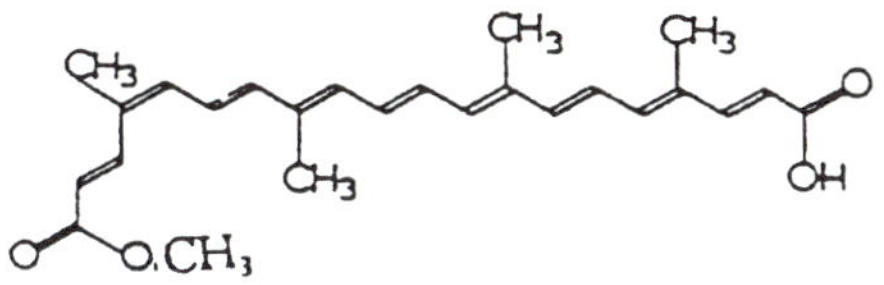

Bixin

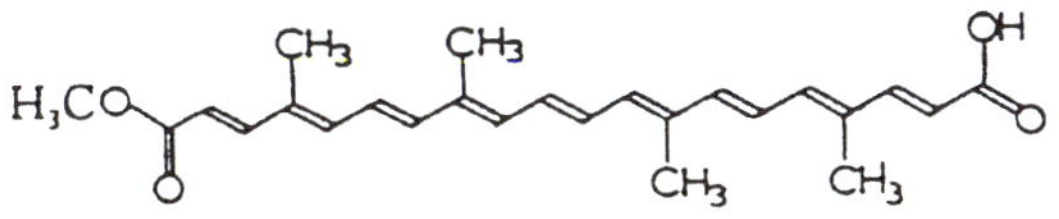

Transbixin

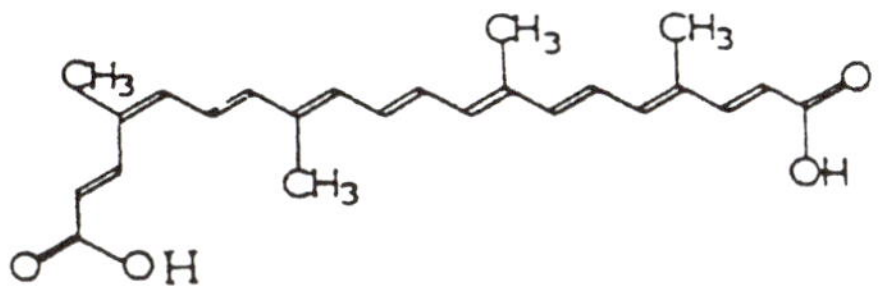

Norbixin

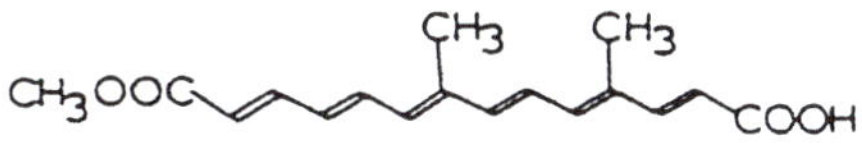

C_{17} yellow pigment

Annatto

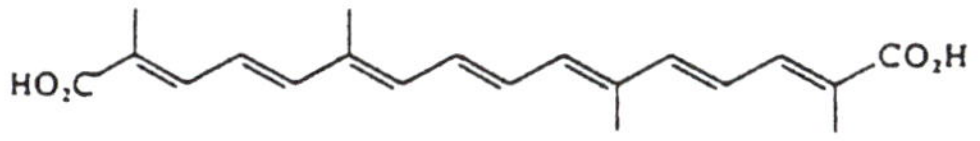

Crocetin

Crocin

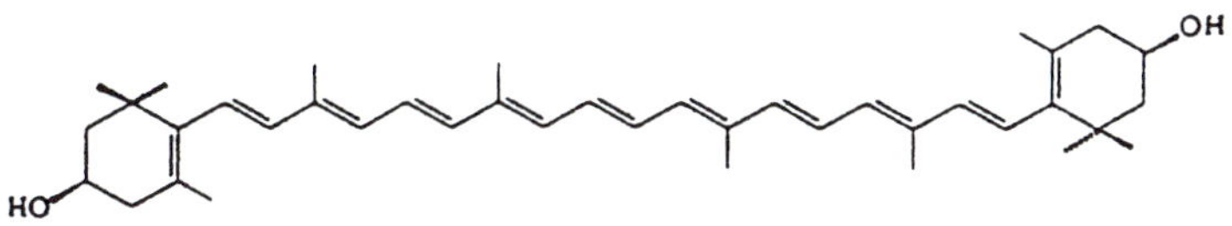

Zeaxanthin

Saffron

Fig. 6-3. The carotenoids of annatto (bixin, transbixin, norbixin, and C_{17} yellow pigment) and saffron (crocetin, crocin, and zeaxanthin).

acid portion of the molecule contributes to solubility in water, and the ester form contributes to oil solubility. This provides flexibility for use in a wide variety of applications. Annatto also contains small quantities of other carotenoids and degradation products of bixin.

Annatto is available in both water-soluble and oil-soluble liquids and powders. The oil-soluble form is somewhat unstable under oxidative conditions; degradation is increased by exposure to light and is catalyzed by metals. Addition of antioxidants, such as ascorbic acid, tocopherols, and polyphenols, helps to minimize oxidative degradation. Annatto shows more stability to exposure to air than other carotenoids and is moderately stable to heat. Little change in color occurs with pH changes, but products with low pH may show a pinkish tinge resulting from isomerization of the pigments.

Saffron

Saffron (CI Natural Yellow 6, CI No. 75100) is a very old colorant dating back to the 23rd century B.C. It has come to be known as the "gourmet spice" because of its high price, but it provides both spice and colorant. Saffron consists of the dried stigmas of the flowers of the crocus bulb, *Crocus sativus,* grown primarily in North Africa, Spain, Switzerland, Austria, Greece, and France. The price is high because it takes about 150,000 flowers to produce 1 kg of colorant.

The pigments in saffron are chemically similar to those in annatto. They are crocetin, a dicarboxylic carotenoid, together with its gentiobioside ester crocin (Fig. 6-3). Gentiobioside is a diglucoside with a β-1-6 linkage. The sugar portion confers solubility in water and makes the colorant very flexible in its applications in a variety of food and pharmaceutical products. The same pigments occur in several other plants, but *Crocus sativus* is the only commercial source. The fruits of the Cape jasmine, *Gardenia jasminoides,* produce the same pigments and have been suggested as an alternate source. But the gardenia fruits supply only the colorant, not the spice flavor, so the sources are quite different. Saffron extracts also contain β-carotene, zeaxanthin, and traces of several other carotenoids. Saffron is an FDA-approved colorant (21 CFR 73), but its use is limited to specialty food products because of its cost.

Saffron preparations are fairly stable to light, oxidation, microbiological attack, and changes in pH. Technically, saffron is a good colorant with high tinctorial strength. Its strength is usually judged by its carotenoid content, as measured by the absorption of an aqueous extract at 440 nm, but Alonso et al (2) suggested that tristimulus methods were more appropriate for classifying samples of saffron.

Paprika

Paprika is a deep red, pungent powder prepared from the ground, dried pods of the sweet pepper *Capsicum annum*. Paprika is produced in many warm countries around the world. Several areas have developed products with specific characteristics, such as the Hungarian paprika and the Spanish paprika. The same peppers are used in salads and as a source of pimiento.

The other principal type of red pepper (also *C. annum*), usually called cayenne pepper or cayenne, is usually much more pungent in flavor. Both types are highly pigmented. The red pepper *C. frutescens* is the source of the highly colored and very pungent Tabasco sauce.

Paprika contains capsanthin and capsorubin (Fig. 6-4), which occur mainly as the lauric acid esters. Smaller quantities of about 20 other carotenoids are also present. Specifications do not attempt to describe the pigment profile but usually specify color strength by the American Spice Trade Association (*ASTA*) color value. This is essentially the absorption at 460 nm in acetone measured against a cobalt solution or a glass standard used as a reference (3).

Paprika oleoresin (EU No. E 160c) is an orange-red oil-soluble extract from *C. annum*. The FDA-approved procedures are specified in 21 CFR Section 73.345. The dried and ground red peppers are extracted with a volatile solvent, followed by removal of the solvent. Chlorinated hydrocarbons were used as solvents in the past, but now hexanes and possibly acetone are used. Paprika and paprika oleoresin add both color and flavor to a product; thus, the applications are usually limited to products in which both characteristics are desirable. For example, the recent rise in demand for tomato products in the form of pizza, salsa, etc., has increased the demand for paprika. Most of the spice extract used in Europe is used to flavor meat products, soups, and sauces. Smaller quantities are used in salad dressings, snacks, processed cheese, confectionery, and baked goods.

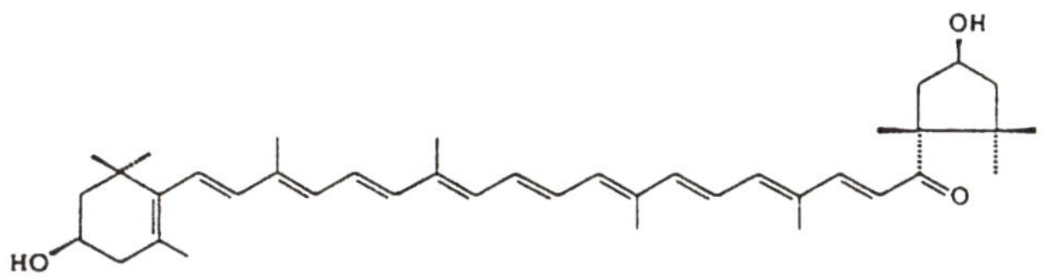

ASTA—American Spice Trade Association

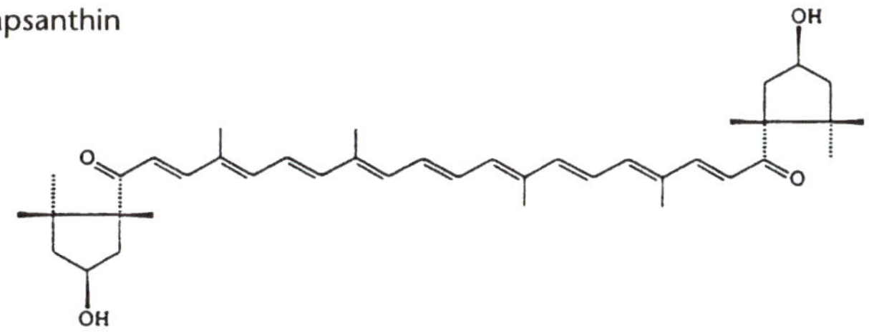

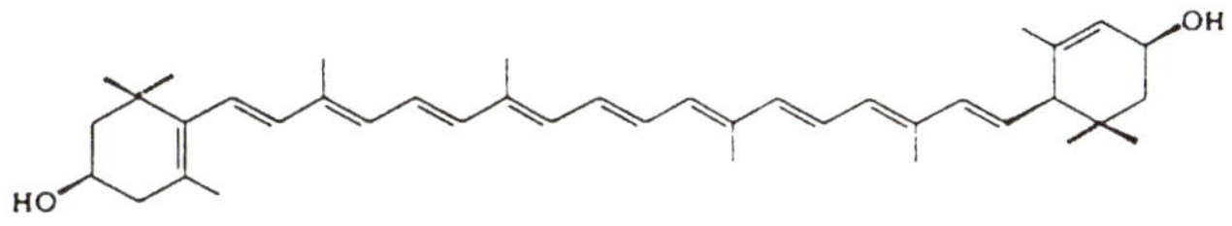

Fig. 6-4. The carotenoids of paprika (capsanthin and capsorubin) and tagetes (lutein).

Tagetes

Tagetes erecta L. (Aztec marigold) is an annual herb that grows 3–4 ft tall in temperate climates. Colorants are available in three forms: dried, ground flower petals; oleoresin extracts; and purified oleoresin extracts. The first two forms are used primarily for poultry feeds and the third for foods, although tagetes extracts are not approved for human food use in the United States. The principal producers are Mexico, Peru, Ecuador, the United States, Spain, and India (4).

The total carotenoid content of tagetes petals is up to 80% lutein (Fig. 6-4), with smaller amounts of zeaxanthin, cryptoxanthin, β-carotene, and about 14 other carotenoids. The lutein compounds exist as dipalmitate, dimyristate, myristate-palmitate, palmitate-stearate, and distearate esters. Because of the seasonality of the crop, the mature marigold flowers are collected and the petals separated prior to storage in large bins containing up to 80 tons. To minimize pigment degradation, the petals are sprayed with an antioxidant, compacted, and covered with a tarpaulin. The petals are removed as needed, pressed, and then dried to less than 10% moisture. They can then be ground and sold as Tagetes meal.

The ground petals can also be extracted with a number of solvents, but hexane seems to be preferred. After removal of the solvent, a brown oleoresin is obtained, which can be incorporated directly into poultry feed. The resin can be further purified by saponification with 40% potassium hydroxide or equivalent alkali and solvent. After removal of the solvent and adjustment of the pH to 6–8, the product can be washed and sold as saponified marigold extract. This product is suitable for incorporation into poultry feed. It can be further purified by washing and being taken up in a suitable vegetable oil or absorbed on calcium silicate, gelatin, starch, etc., to produce a dry powder suitable for use as a food colorant. Enhanced stability is claimed when the esters are partially saponified, then neutralized to pH >8 with acetic, propionic, or lauric acids. The final product usually contains 10–20% by weight of the original xanthophyll esters.

Colorants from Tagetes are available in a variety of forms. Formulations for animal feeds are usually ground dried petals, oleoresins, or saponified oleoresins. Food colorants are available in a number of forms, such as purified lutein esters in oil-soluble or water-dispersible systems, spray-dried emulsions, gum-based emulsions, and emulsifier-based emulsions. They show good stability to heat, light, pH changes, and sulfur dioxide. They are susceptible to oxidation, which can be minimized through encapsulation or the addition of antioxidants such as ethoxyquin, ascorbic acid, tocopherols, or butylated hydroxyanisole and butylated hydroxytoluene. The strength of Tagetes extracts is usually measured as the absorption of a 1% solution in a 1-cm cell in an appropriate solvent and is often quantified by calculations using the specific extinction coefficient of lutein or lutein esters.

Tagetes colorants provide yellow to orange colors suitable for use in pastas, vegetable oils, margarine, mayonnaise, salad dressings, baked goods, confectionery, dairy products, ice cream, yogurts, citrus juices, and mustard. However, in the United States, Tagetes meal and its extracts are approved only as colorants in poultry feed.

Lycopene

Lycopene (Fig. 6-1) occurs as the major (85–90%) pigment in red tomatoes, *Lycopersicon esculentum*. The other pigments are β-carotene

(10–15%) and small quantities of about 10 other carotenoids. In spite of the fact that large quantities of lycopene are available in the waste from the tomato processing industry, colorants containing lycopene were not commercially available in the past. This was probably due to the belief that lycopene was susceptible to degradation by oxidation and light. Recently, however, a combination of better manufacturing practices and the development of a tomato cultivar particularly high in lycopene led to the commercialization of lycopene as a food colorant (5).

Preparation of lycopene extracts from tomatoes is relatively simple, involving an alkali saponification and extraction with a mixture of solvents such as acetone and hexane. The acetone can be removed by washing with water and the hexane by vacuum treatment, leaving a relatively pure lycopene extract, which nevertheless does contain the other carotenoids in the tomato. Commercialization probably was also helped by the health aspects. Lycopene is the carotenoid present in human blood plasma in the highest concentration (5) and is believed to be an efficient *in vivo* radical scavenger. Lycopene preparations are being marketed as nutraceuticals and as food colorants. Colorant preparations containing lycopene are currently not allowed in the United States.

Miscellaneous Carotenoid Extracts

The red oil obtained from the fruit of the palm tree *Elaeis guineensis* is very highly pigmented, with about 500 mg of carotenoids per kilogram of oil. The carotenoid mixture is very complex, containing mainly β-carotene as well as about 20 other carotenoids. The oil can be used as an ingredient to color margarine and other oil-based products, but its chief use is as a cooking oil because of its distinctive flavor or, after refining, as a general purpose edible oil. Carotenoids are also extracted from the red oil to form a concentrated carotenoid containing primarily β- and α-carotene (Fig. 6-5). This is approved for use as a colorant in the Europe but currently not in the United States. It is used in the United States for its nutritional value.

Xanthophyll pastes, well known in Europe, consist of extracts of alfalfa (lucerne), nettles, or broccoli. Unless saponified, they are green because of their high content of chlorophyll. Many concentrated xanthophyll pastes contain as much as 30% carotene, with the major pigments being lutein (45%), β-carotene (25%), neoxanthin (15%), violaxanthin (15%), and a number of minor pigments (Fig. 6-5). Highly concentrated extracts from alfalfa became available some years ago as by-products of the Pro-Xan process for preparing protein concentrates from alfalfa. The carotenoid extracts were widely promoted as colorant additives for poultry feed, but the Pro-Xan process was never commercially successful.

Extracts from citrus peels have been suggested as a means of augmenting the natural color of orange juice, since the more highly colored juices command a premium price, but addition of colorants

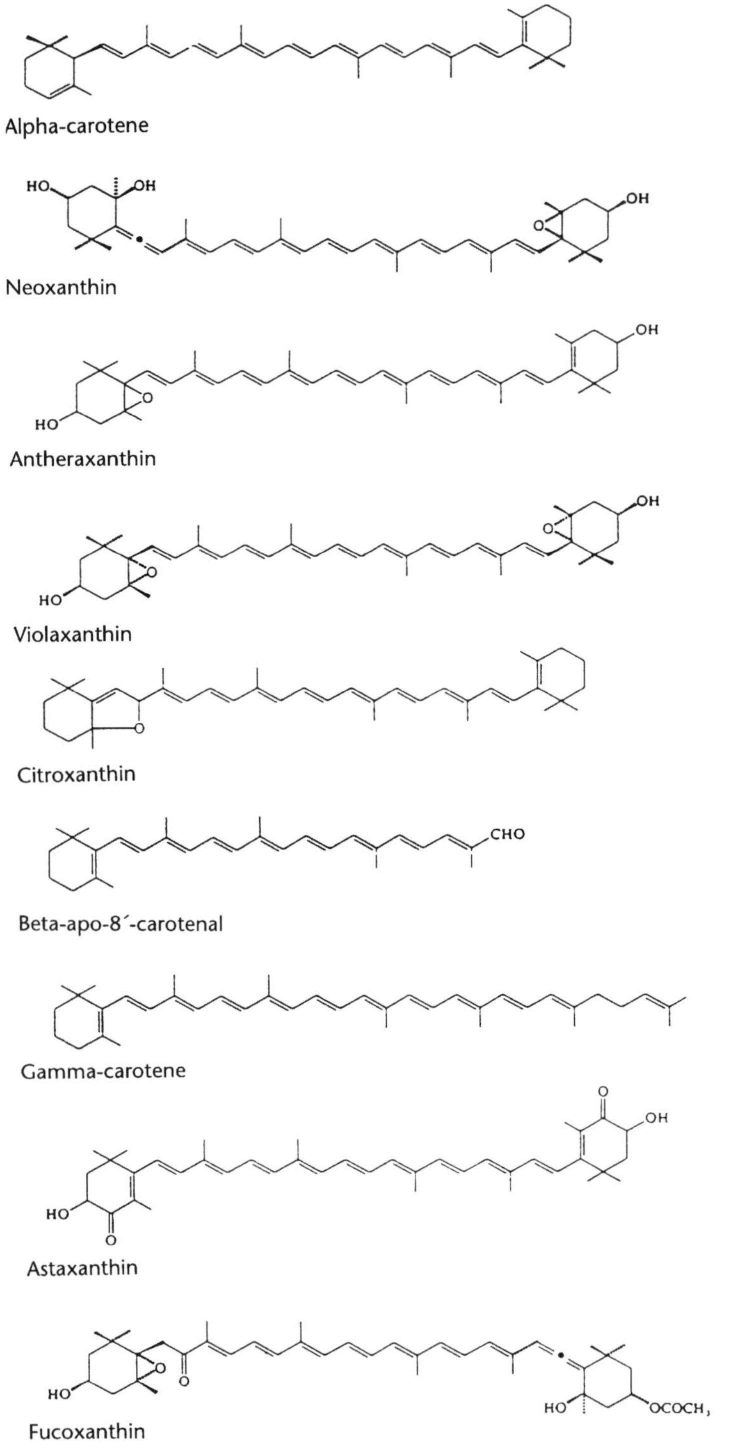

Fig. 6-5. Selected carotenoid structures from miscellaneous sources.

from citrus peel is not permitted in the United States at the present time. About 115 carotenoid pigments have been reported in citrus. Citrus peel extracts contain β-carotene, cryptoxanthin, antheraxanthin, violaxanthin, lycopene, citroxanthin, and β-apo-8′-carotenal (Fig. 6-5) together with a number of others.

Colorants from carrots usually contain about 80% β-carotene and up to 20% α-carotene, plus small amounts of γ-carotene and several other minor carotenoids. Some of the high-pigment strains of carrots used for colorant extracts also contain lycopene.

Astaxanthin (Fig. 6-5) is a desirable addition to the diet of salmon and trout in aquaculture because of its ability to impart a desirable red color to the flesh. The usual sources of astaxanthin are the by-products of the lobster and shrimp processing industry, but the demand exceeds the supply. This has led to an interest in growing the red yeast *Phaffia rhodozyma* as a raw material for a concentrated extract. Unfortunately, *Phaffia* produces the wrong optical isomer of astaxanthin for optimal accumulation in the flesh of salmon, but it is the only yeast known to produce astaxanthin, and one process has been patented (6). Hinostroza et al (7) reported that carotenoids from three sources, synthetic astaxanthin, crabs (*Pleuroncodes planiples*), and *Phaffia rhodozyma*, were deposited in trout muscle at the rate of 7.8, 5.0, and 4.0%, respectively. Astaxanthin is permitted for addition to fish food. There is ongoing interest in developing other plant sources to compete with the nature-alike sources described in the next section.

Several mutants of the carotogenic molds *Blakeslea trispora* and *Phycomyces blakesleeanus* produce high concentrations of β-carotene, but recent interest has shifted to microalgae. Species of *Dunaliella* can accumulate up to 10%

β-carotene (dry weight). Ponds in Australia originally developed to produce salt by evaporation of salt water are now being used to grow *Dunaliella*. Other installations exist in Hawaii. Contamination in the growing phase is minimized by the high salt concentration, but apparently purification of the extracts is difficult. Regardless of the difficulties, both sources command a premium because of their "natural" association. Interestingly, the same ponds can be used to grow *Haematococcus* species, which accumulate astaxanthin. In view of the widespread occurrence of the carotenoid pigments, it is not surprising that other plants and animals would be suggested as potential sources of colorants. These include krill, chlorella, shrimp, algae, bacteria, molds, and a variety of plant sources (8).

Synthetic Carotenoids

Plant extracts of carotenoids have been used for centuries as food and cosmetic colorants, so it was only natural that synthetic carotenoids would become available as colorants. The original synthesis of β-carotene was reported in 1950, and commercial production followed in 1954. It was followed by β-apo-8′-carotenal in 1962 and canthaxanthin in 1964. The methyl and ethyl esters β-apo-8′-carotenoic acid and citranaxanthin (Fig. 6-6) followed. β-Carotene, canthaxanthin, and apo-carotenal are the only synthetics currently permitted in the United States. Zeaxanthin and astaxanthin are under development at present. Astaxanthin was recently approved as a feed additive for coloration of fish.

By virtue of their structure, all carotenoids are susceptible to degradation by oxidation, light, and heat. In addition, they are insoluble in water and almost insoluble in oil. The challenge in overcoming these problems reads like a textbook case for food formulation. Nevertheless, commercial preparations are currently available for almost any type of food in the yellow to red range.

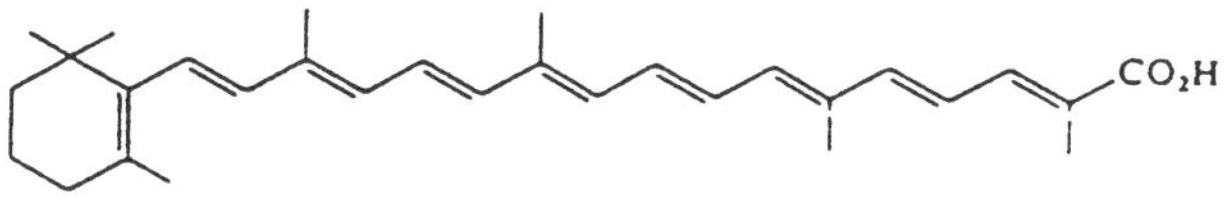

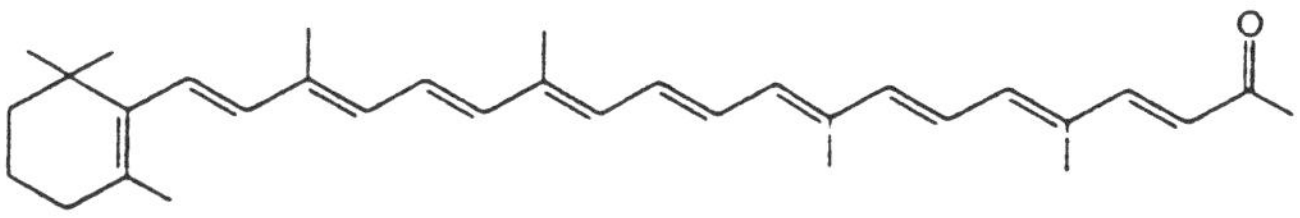

Fig. 6-6. Carotenoids synthesized for commercial colorant applications. β-Apo-8′-carotenoic acid is usually in the form of the methyl or ethyl esters. However, β-carotene, canthaxanthin, and apo-carotenal are the only synthetics permitted for use in the United States.

A variety of approaches is used to prepare appropriate formulations. These involve reducing particle size for suspensions in oil and stabilization with antioxidants. Three approaches are used to develop water-dispersible products: formulation of colloidal suspensions, emulsification of oily solutions, and dispersion in suitable colloids. These products can be combined with a wide variety of additives in the protein, carbohydrate, and lipid categories and stabilized with antioxidants (9). Carotenoid colorants are appropriate for margarines, oils, fats, shortenings, fruit juices, beverages, dry soups, canned soups, dairy products, milk substitutes, coffee whiteners, dessert mixes, preserves, syrups, confectionery, salad dressings, meat products, pasta, egg products, baked goods, and many others.

Health Aspects

In the United States in the 1960s and 1970s, food additives received more than their share of criticism. The colorants were an easy target for the "junk food" concept since they were seen to be of cosmetic value only. Who would accept any risk at all for only cosmetic reasons? The carotenoids were looked on more favorably because of their well-known association with vitamin A and vitamin A precursors.

There is good evidence that antioxidants such as β-carotene are effective against three major health problems: cancer, coronary artery diseases, and macular degeneration of the eye. However, results from two large recent studies have shown that β-carotene in capsule form is ineffective in protecting smokers from lung cancer. The protective mechanism of antioxidants is believed to involve minimizing the effect of reactive oxygen species. The actual carotenoid compounds involved are not known at the present time, but the range of carotenoid compounds involved as antioxidants is likely to be much broader than those involved as provitamin A. This has led to a movement to develop a recommended dietary allowance for carotenoids in addition to that for vitamin A.

Currently, food colorant preparations are optimized for colorant potential, not for health potential. This may change. One commercial preparation emphasizes both the colorant properties and the health aspects. The much wider profile of carotenoid pigments in natural extracts would appear to give the extracts an advantage over synthetic compounds. However, when the carotenoids involved in health benefits other than vitamin A become better understood, there is nothing to prevent manufacturers from synthesizing carotenoids in addition to the six currently produced for the colorant market. An additive with a positive health image is very desirable.

References

1. Bauernfeind, J. C. 1981. *Carotenoids as Colorants and Vitamin A Precursors*. Academic Press, New York.
2. Alonso, G. L., Sanchez, M. A., Salinas, M. R., and Navarro, F. 1996. Color analysis of saffron. In: *Proceedings of the Second International Symposium on Natural Colorants*. P. Hereld, Ed. The Hereld Organization, Hamden, CT.
3. Marmion, D. M. 1991. *Handbook of U.S. Colorants*, 3rd ed. John Wiley & Sons, New York.
4. Verghese, J. 1996. Focus on xanthophylls from *Tagetes erecta* L.—The giant natural colour complex. *Proceedings of the Second International Symposium on Natural Colorants*. P. Hereld, Ed. The Hereld Organization, Hamden, CT.
5. Nir, Z., and Hartal, D. 1996. Lycopene in functional foods. *Proceedings of the Second International Symposium on Natural Colorants*. P. Hereld, Ed. The Hereld Organization, Hamden, CT.

6. Fenoe, B., Christenson, I., and Larson, R. 1987. Astaxanthin-producing yeast, astaxanthin isolation and its use in animal feed. U.S. patent 5,709,056.
7. Hinostroza, G. C., Huberman, A., Espino, G. de la Lanza, and Monroy, H. 1996. Pigmentation of rainbow trout *Oncorhynchus mykiss* by astaxanthin from red crab *Pleuroncodes planipes* in comparison with synthetic astaxanthin and *Phaffia rhodozyma* yeast. *Proceedings of the Second International Symposium on Natural Colorants*. P. Hereld, Ed. The Hereld Organization, Hamden, CT.
8. Francis, F. J. 1986. *Handbook of Food Colorant Patents*. Food and Nutrition Press, Westport, CT.
9. Francis, F. J. 1995. Carotenoids as colorants. World of Ingredients. Sept/Oct. 34-38.

CHAPTER 7

Anthocyanins and Betalains

Anthocyanins

The anthocyanins are probably the best known of the natural pigments. Ubiquitous in the plant kingdom, they are responsible for many of the orange, red, blue, violet, and magenta colors in plants. Their very visibility, combined with their role as taxonomic markers, has attracted the efforts of many research workers in the past 75 years. As colorants, they go back to antiquity—the Romans used highly colored berries to augment the color of wine.

In view of the ubiquity of the anthocyanins and their high tinctorial power, it comes as no surprise that many plant sources have been suggested as colorants. Francis listed pigment profiles and methods of extraction for over 40 plants as potential sources (1,2) and also listed 49 patents on anthocyanin sources (3). However, despite the large number of potential sources, only two, grapes and red cabbage, have had commercial success. More recently, elderberry, aronia (chokeberry), and black carrot have been moderately successful.

Colorants from grapes have been available for nearly 120 years, primarily from press cake as a by-product of the wine industry. Grapes are the world's largest fruit crop for processing. In 1995, the annual production of grapes was estimated to be 60 million metric tons, of which about 80% was used for making wine (4). This ensures a limitless source of inexpensive raw material for colorant production.

CHEMICAL COMPOSITION

The composition of any colorant containing anthocyanins reflects the composition of the raw material. At present that material is mostly grapes and red cabbage, but other sources may become available in the future. The approximately 275 anthocyanins known at present are part of about 5,000 *flavonoid* compounds (5) of similar chemical structure, as reported in the literature in 1988. "Flavonoid" is a general term describing a large group of polyphenolic compounds. Anthocyanins are also polyphenols and are usually classified within the flavonoid group. The only difference is that the anthocyanins are usually orange to red, whereas the other flavonoids are usually colorless or yellow.

Anthocyanins are composed of an *aglycone* (anthocyanidin), sugar, and perhaps organic acids. Twenty-two aglycones are known, of which 18 occur naturally. Only six (pelargonidin, cyanidin, delphini-

---

In This Chapter:

Anthocyanins
  Chemical Composition
  Commercial Preparation
  Synthetic Compounds
  Considerations in
    Commercial
    Applications
  Health Effects

Betalains
  Chemistry
  Occurrence
  Commercial Preparation
  Applications

Flavonoid—A general term for a very large group of yellow to red pigments found in many plants.

Aglycone—The component in a flavonoid pigment (one of two or more compounds in the same molecule) that contributes most of the color.

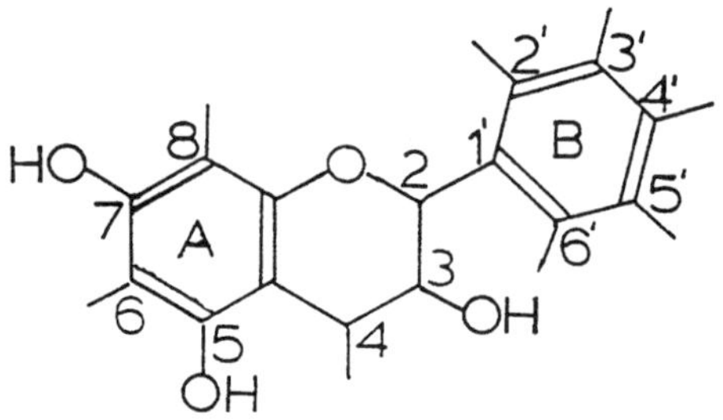

Fig. 7-1. The anthocyanidin nucleus. For pelargonidin, 4´ = OH. For cyanidin, 3´, 4´ = OH. For delphinidin, 3´, 4´, 5´ = OH. For peonidin, 4´ = OH, 3´ = OMe. For petunidin, 4´, 5´ = OH, 3´ = OMe. For malvidin, 4´ = OH, 3´, 5´ = OMe. All other substitution positions = H.

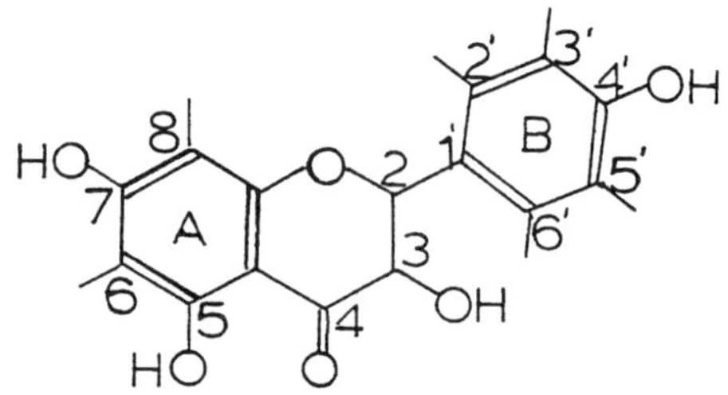

Fig. 7-2. The flavonol nucleus. For kaempferol, no substitutions. For quercetin, 3´ = OH. For isorhamnetin, 3´ = OMe. For myricetin, 3´, 5´ = OH. All other substitution positions = H.

Enocyanin—A general term for a colorant made from grape skins or other by-products of the wine or grape juice industry.

din, peonidin, petunidin, and malvidin) are important in foods; the structure of these is shown in Figure 7-1. Free aglycones occur very rarely in plants, as they are nearly always combined with sugars. One reason for this is that the sugars stabilize the molecule. In order of relative abundance, the sugars found with aglycones are glucose, rhamnose, galactose, xylose, arabinose, and glucuronic acid.

Anthocyanins may also be acylated, which accounts for the third component of the molecule. One or more molecules of the acyl acids p-coumaric, ferulic, and caffeic or the aliphatic acids malonic and acetic may be esterified to the sugar molecule. The aglycones in grapes are cyanidin, peonidin, malvidin, petunidin, and delphinidin, and the organic acids are acetic, coumaric, and caffeic. The only sugar present is glucose. Grapes usually have a very complex anthocyanin profile; the Concord variety has 31 anthocyanins, the greatest number in any single variety. The 15 anthocyanins in red cabbage contain only cyanidin and glucose, along with the acids ferulic and coumaric.

Plants that contain anthocyanins also invariably contain flavonoids. The distribution of flavonoids is more widespread than that of anthocyanins. Although the number of flavonoids is large (5,000), the pigment profile of a plant is usually characteristic of its plant family. The flavonoids of grapes include the 3-glucoside, 3-galactoside, 3-glucuronide, and 3-glucosyl-arabinoside of kaempferol; the 3-glucoside, 3-glucuronide, 3-rutinoside, 3-glucosyl-galactoside, and 3-glucosyl-xyloside of quercetin; myricetin; myricetin-3-glucuronide; isorhamnetin-3-glucoside; dihydroquercetin-3-rhamnoside; and dihydrokaempferol-3-rhamnoside (6). The structure of the flavonol aglycones is shown in Figure 7-2. Several other polyphenolic compounds are also present in grapes, such as (+)-catechin, (–)-epicatechin, (+)-gallocatechin, (–)-epigallocatechin, epicatechin-gallate, catechin-gallate, catechin-catechin-gallate, and a variety of tannins.

Many phenolic acids are also present in grapes, such as p-hydroxybenzoic, salicylic, gallic, cinnamic, p-coumaryl tartaric (coutaric), caffeyl tartaric (caftaric), feruloyl tartrate, p-coumaryl-glucoside, and feruloyl glucoside. The polyphenolic compounds in grapes, in view of their importance in wines, have received much more research attention than those in red cabbage, but the latter can be assumed to also have a complex polyphenol profile. A typical compound from each group of compounds is shown in Figure 7-3.

The composition of grape extracts is further complicated by polymers produced by association with catechins and flavonoids with or without interaction with acetaldehyde. Actually, most of the tinctorial power in *enocyanin* preparations is a result of the complexed anthocyanins. The overall conclusion is that it is impossible to specify accurately the composition of enocyanin preparations. Commercial specifications are usually confined to tinctorial power, total acidity, percent solids, percent ash, heavy metals, sulfur dioxide, tannins, and alcohol. The tinctorial strength of grape extracts is often expressed as the absorbance of a 1% solution in a 1-cm cell at 520 nm in citrate buffer at pH 3.0. The tinctorial power of red cabbage extracts is sometimes expressed as the absorbance of a 10% solution in a 1-cm cell at 520 nm measured in citrate buffer at pH 3.0 (e.g., E 10%, 1 cm = 190).

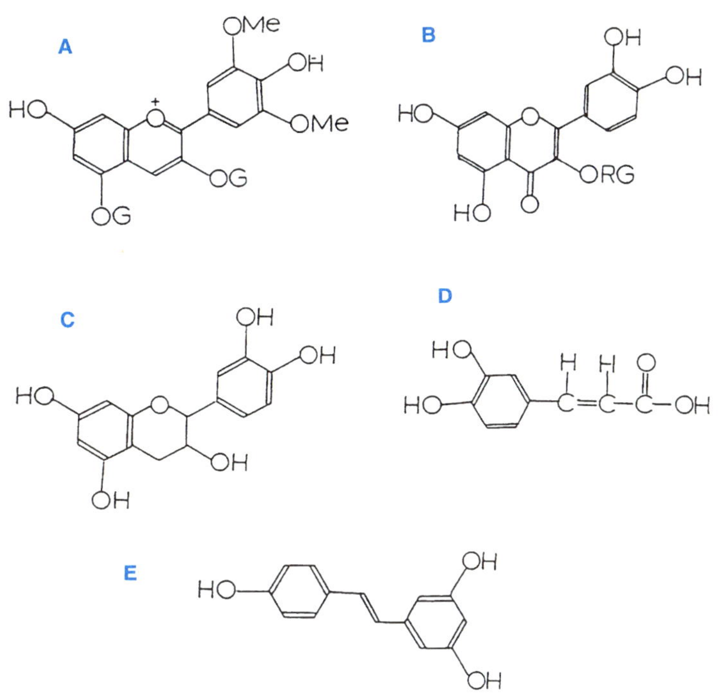

COMMERCIAL PREPARATION

The most common commercial method of producing Grape Skin Extract involves treatment of the skins with water containing up to 3,000 ppm of sulfur dioxide or its equivalent in bisulfite or metabisulfite. After 48–72 hr, the liquid is removed from the skins, filtered, desulfured, and concentrated. Sometimes a fermentation step is allowed before the extraction, in which case the alcohol is removed during subsequent processing. The final product is a particle-free liquid of high tinctorial power, which may also be dried to produce a water-soluble dry powder.

Fig. 7-3. Typical components of grape colorant extracts: a red anthocyanin, malvidin-3,5-diglucoside (A); a yellow flavonoid, quercetin-3-rhamnoglucoside, (rutin) (B); a component of tannin, catechin (C); a phenolic acid, caffeic acid (D); a stilbene phytoalexin, resveritol (E).

The presence of sulfur dioxide in the extracting medium results in increased extraction of the pigments as well as increased stability of the final product. The sulfite-anthocyanin complex is colorless, so the sulfite must be removed before the final concentration and filtration step. Addition of acid makes the extraction step more efficient, but addition of mineral acid requires neutralization at some point. If tartaric acid is used, it can be removed by addition of potassium hydroxide since the resulting precipitate of potassium hydrogen tartrate can be easily removed by filtration.

Purification of the crude extract can be accomplished by passing the extract through an ion-exchange resin bed with the added advantage that the colorant eluate can be fractionated. The first portion of the eluate is redder than the bluer trailing portion because the higher molecular weight compounds go through the resin bed at a slower rate. Treatment with resins produces a purer product but at a higher price. The FDA-approved procedures for extraction are specified in 21 CFR 73.170.

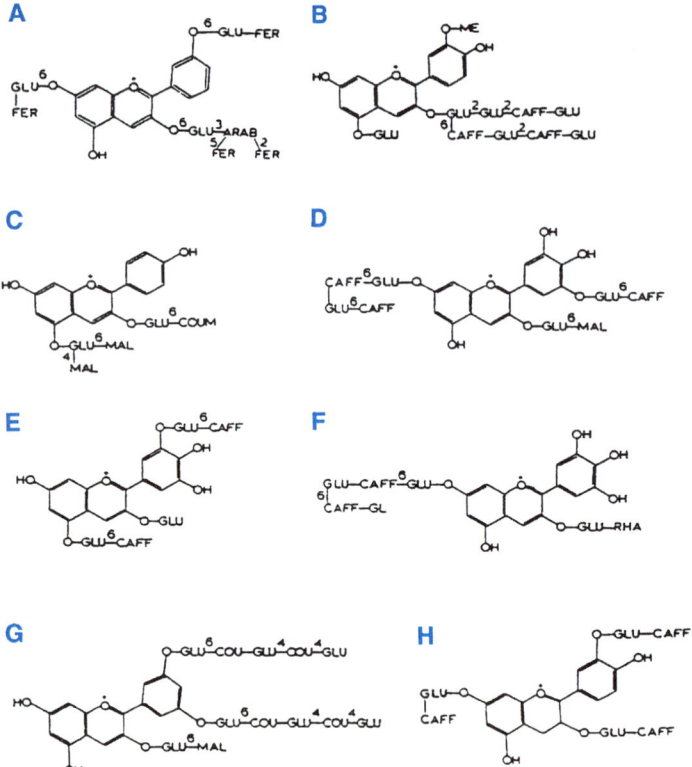

Fig. 7-4. Structures of selected acylated anthocyanins. A, anthocyanin from *Tradescantia pallida*; B, anthocyanin from the cultivar Heavenly Blue morning glory, *Ipomoea tricolor*; C, monardaein from *Monarda didyma*; D, cinerarin from *Scenecio cruentus*; E, gentiodelphin from *Gentiana makinoi*; F, platyconin from *Playcodon grandiflorum*; G, ternatin D from *Clitoria ternatea*; H, zebrinin from *Zebrina pendula*. Glu and Rha refer to the sugars glucose and rhamnose. Caff, Coum, Fer, and Mal refer to the acyl acids caffeic, coumaric, ferulic, and malonic. The superscripts refer to the carbon linkages of the sugar acids.

Extraction of grape skins is much more efficient with acidified methanol or ethanol, but this requires another step to remove the alcohol. Wineries are usually familiar with handling alcohol, but other potential colorant producers seem reluctant to put an alcohol recovery unit in a food plant. As a result, alcohol extraction does not seem to have commercial appeal.

Current emphasis is on attempts to make colorants containing anthocyanins more stable. Considerable research has emphasized the association of anthocyanins with themselves or with a series of other compounds such as flavonoids, polysaccharides, proteins, tannins, and other polyphenolic compounds. This association, termed "intermolecular copigmentation," has been less effective than association at the intramolecular level, which usually involves acylation of the molecule. The acyl portions of the molecule are believed to stack above and below the point of attack near the C-2 position (Fig. 7-1), thus providing a more stable molecule. The co-pigmentation effect is strongest with the flavonols quercitin and rutin and the C-glycosyl flavones such as swertisin (7).

Intermolecular reactions may be important in stabilizing the colors of plant tissues *in vivo* but have had little effect in stabilization of food colorants. Production of more stable colorants may have to come from the use of more stable pigments. It is well established that acylated compounds are more stable than their nonacylated counterparts (8). The most stable pigments reported in the literature are all highly acylated (Fig. 7-4).

Highly acylated compounds also have another feature (9) that makes them desirable as colorants: they usually have an extra absorption band at 560–600 and/or 600–640 nm in weakly acid or neutral solutions. This means that the compounds are highly colored at pH values above 4.0, where conventional anthocyanins would be nearly colorless. The two extra bands result from the tendency of anthocyanins to form quinoidal base structures in weakly acid or neutral solutions. Two forms may be present: the 7-keto form and the 4'-keto form. The 7-keto form predominates in conventional anthocyanins, where only the normal 520–560 nm band appears. If the 7-OH position is blocked by methylation or glycosylation, the

4'-keto form appears and the extra band is formed. Figure 7-5 shows the absorption spectra of anthocyanin from a plant, *Tradescantia pallida*, at two pH values.

Several sources of acylated anthocyanins have been patented as potential food colorants (3). Among these, only red cabbage and red sweet potatoes are normal food sources. The structures of the highly acylated anthocyanins are complex, making commercial synthesis unlikely. Production of this group of pigments, if it does occur, is likely to be through a tissue culture approach. Obviously, such pigments would also require regulatory approval.

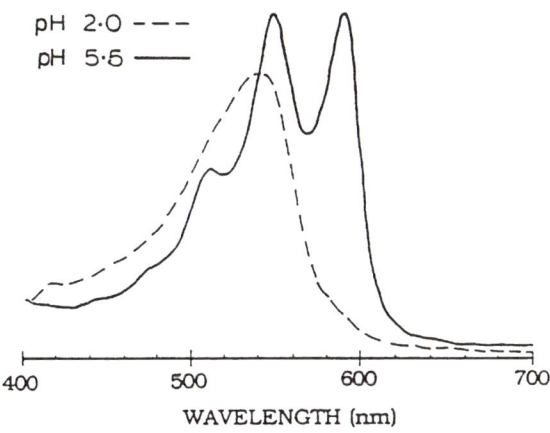

Fig. 7-5. Absorption spectra of the major anthocyanin in *Tradescantia pallida* at pH values 2.0 and 5.5. Reprinted, with permission, from (9).

SYNTHETIC COMPOUNDS

Studies on synthetic compounds related to the anthocyanins have shown that, with methyl or phenyl moieties substituted at the C-4 position in the molecule, these compounds are virtually resistant to nucleophilic attack at the C-2 position. This observation led to the suggestion that a number of synthetic analogues would prove to be very stable (10). The first reported natural anthocyanin with substituted C-4 was purpurinidin fructoglucoside, isolated from willow bark, but a shortage of raw material precluded its commercial development (11). A second group of C-4-substituted anthocyanins, derived from anthocyanins in red wine, was reported (12). The new pigments, containing a vinylphenol group attached to positions 4 and 5 of the grape anthocyanins, were named anthocyanin-3-glucoside adducts. In a study (13), researchers prepared the vinylphenol adducts of five of the major anthocyanins of wine, namely delphinidin-3-glucoside, cyanidin-3-glucoside, petunidin-3-glucoside, peonidin-3-glucoside, and malvidin-3-glucoside, and tested their stability at 55°C in a model beverage at pH 3 and 5. The adduct pigments were much more stable than the original pigments, particularly at pH 5. The adducts also were unaffected by SO_2. The color of the adducts was slightly more orange than that of the corresponding anthocyanins. The authors did not state the concentration of the adducts found in red wine. Chemical synthesis of a number of C-4-substituted analogues of anthocyanins is feasible but obviously would require testing for safety and approval from a regulatory agency.

CONSIDERATIONS IN COMMERCIAL APPLICATIONS

In view of the large body of research on the appearance, stability, analysis, and plant breeding of anthocyanins naturally occurring in fruit and vegetable products, it is not surprising that anthocyanin colorants would be used to enhance the aesthetic appeal of existing plant products or formulated substitutes. Actually, one of the first applications of enocyanin was to improve the color of wine. In the

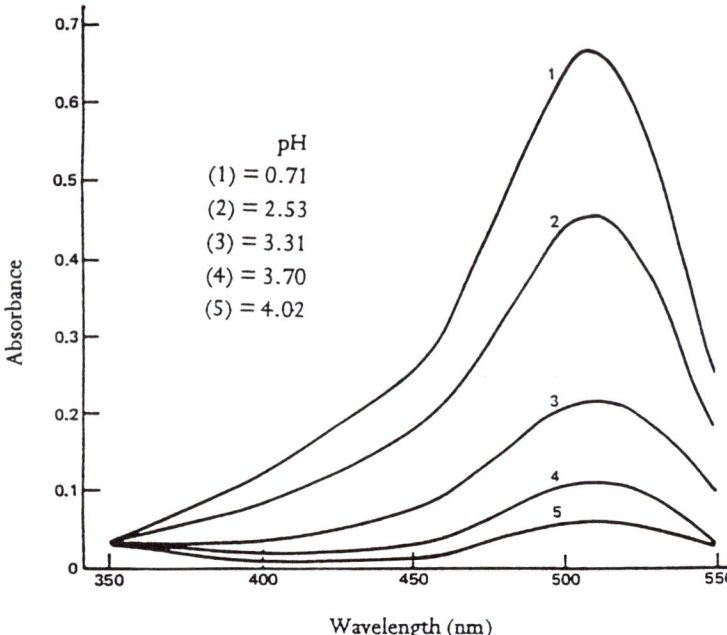

Fig. 7-6. Absorption spectra of cyanidin-3-rhamnoglucoside in buffer solutions at pH values 0.71–4.02. The concentration of the pigment is 1.6 × 10² g/L. Reprinted, with permission, from (14).

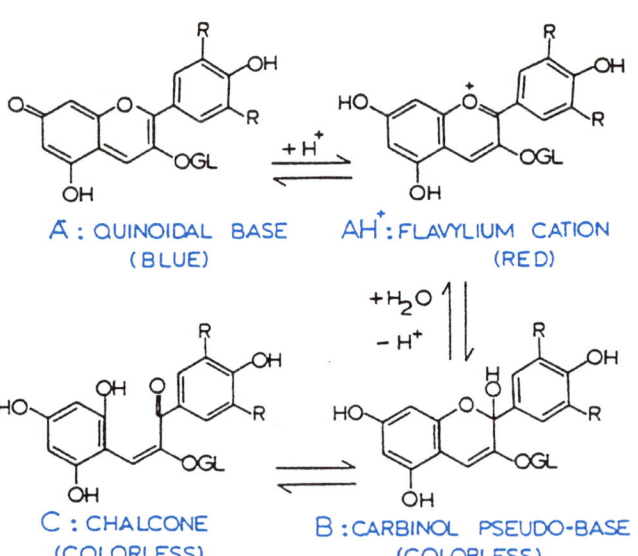

Fig. 7-7. Structural changes in anthocyanins with pH. The example is malvidin-3-glucoside at 25°C.

United States, fruit drinks are the biggest market. Colorants containing anthocyanins have been suggested for beverages, jellies, jams, ice cream, yogurt, gelatin desserts, canned fruits, fruit sauces, candy and confections, bakery fillings, toppings, drink-mix crystals, pastries, cosmetics, and pharmaceuticals.

The color intensity of anthocyanins changes with pH, with the greatest intensity at pH values less than 4.0. Figure 7-6 shows the changes in absorption of cyanidin-3-rhamnoglucoside as the pH is changed from 0.71 to 4.02. Figure 7-7 shows the chemical changes that accompany a pH change, and Figure 7-8 shows the amounts of each pigment present at each pH value. It is possible to use an anthocyanin colorant to color a product with a pH value above 4.0, but the color is bluish and considerably more colorant is required. The reduction in color intensity as the pH is raised is known as the pH effect. It is much more obvious with preparations that contain a high proportion of monomeric anthocyanins and is less obvious in preparations with a higher content of polymerized or degraded pigments. However, all preparations show the pH effect to some degree.

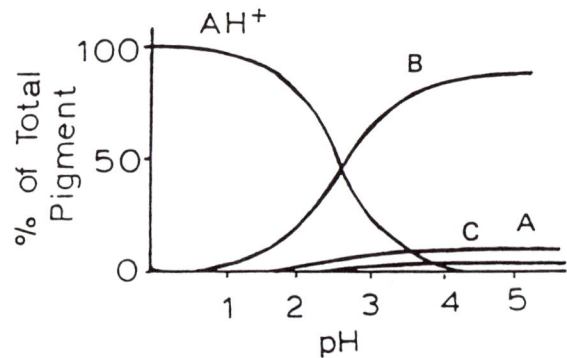

Fig. 7-8. Effect of pH on the distribution of anthocyanin structures of malvidin-3-glucoside. A, B, C, and AH⁺ refer to the forms in Figure 7-5.

In addition to pH sensitivity, anthocyanins are susceptible to degradation by light, heat, oxygen, iron, copper, tin, ascorbic acid, and sulfur dioxide. The product developer needs to take these factors into consideration. Despite these apparent problems with stability, this group of colorants has had good commercial success.

HEALTH EFFECTS

The complexity of the grape colorants is a mixed blessing. From a purely colorant point of view, it is of little importance provided that the product is technologically efficient and safe. After 120 years of use, there is little question about safety, but the attribution of health effects is another matter. Health effects have been attributed to anthocyanins and flavonoids for a long time and may have been the inspiration for the term "bioflavonoids." A growing body of data supports the claim that polyphenols are very good oxygen scavengers and by this mechanism reduce a number of serious health situations (15). Interest in the antioxidative potency of wines was piqued by the "French paradox," in which the mortality from cardiovascular heart diseases was lower than the level predicted from the intake of dietary saturated fatty acids. The beneficial effects were greater in association with alcohol taken in the form of wine, suggesting that there may be a protective effect from other components of wine. There is some evidence that the antioxidative effects may be responsible for reduction in lipid peroxidation, atherosclerosis, and thrombosis. Resveritrol, a stilbene phytoalexin produced in plants as a reaction to stress, has been in the news lately in connection with benefits for cardiac function. By the same reasoning, it has been suggested to be an anticarcinogen.

In addition to the antioxidant effects, a wide variety of beneficial effects attributed to anthocyanins (16) has led to their use as bacteriostatics (antivirals, cancer prevention), eyesight enhancers (night blindness hinderers), drug efficiency improvers, hypoallergenic cosmetics, and neutraceuticals. While there is a large amount of data, anecdotal and otherwise, on the beneficial effects of polyphenols, including anthocyanins, the actual causative agents may be difficult to identify.

In the future, it may be possible to formulate colorant preparations to optimize both the technology of coloration and the attributes to health. However, we are a long way from that ideal goal at present, and current practice is to optimize their properties as a colorant.

Betalains

The red roots of beet, *Beta vulgaris,* have been known for centuries as an attractive food and as a means of imparting a desirable red color to other foods. Extracts of red beets as colorants are a relatively recent development, but the concept dates back well over 100 years. The red pigment from beets is similar to that in the berries of pokeweed,

Phytolacca americana, which are intensely colored and have provided some insurance for the wine industry against years with poorly colored grapes. The addition of pokeberry juice to wine was forbidden in France in 1892, primarily because the juice also contains a purgatory and emetic saponin called phytolaccatoxin. The red pigment was formerly called "phytolaccanin" until research established that it was identical to betanin in beets (17,18). The same researchers also published a method to remove the phytolaccatoxin from pokeberry juice.

In the 1980s, considerable interest was shown in the United States in the development of colorants from beets, but unfortunately, it coincided with the unfolding of the role of nitrites in the formation of toxic nitrosamines. Beets are notorious accumulators of nitrates and nitrites during the growing period, and the necessity of reducing the nitrite level was one more hurdle to overcome. Extracts from beets are the only betalain colorants permitted in the United States today.

CHEMISTRY

The betalain group contains about 50 red pigments called betacyanins and 20 yellow pigments termed betaxanthins. The betacyanins are all derivatives of the diazoheptamethin structure shown in Figure 7-9. The color is due to the resonating structures shown in Figure 7-10. If the R or R' group extends the resonance, the pigment is red, as in betanin (Fig. 7-11). If the resonance is not extended, the pigment is yellow, as in vulgaxanthin (Fig. 7-12). The betacyanins contain the aglycones betanidin and isobetanidin and the rarer form 2-decarboxybetanidin (Fig. 7-13). Another rare form, neobetanidin, may also be found. The aglycones are usually combined with glucose and much less frequently with sophorose and rhamnose. The betacyanin molecule may also be acylated with sulfuric, malonic, 3-hydroxy-3-methylglutaric, citric, *p*-coumaric, ferulic, caffeic, or synapic acids. Substitution on the R group of betaxanthin includes residues from glutamic acid to produce vulgaxanthins 1 and 2 (Fig. 7-12). Other betaxanthins are formed when the R' group (Fig. 7-9) includes methionine sulfoxide, tyramine, dihydroxyphenylalanine, or 5-hydroxynorvaline. The R group in Figure 7-12 for betaxanthins is usually hydrogen. Another important betaxanthin called indicaxanthin (Fig. 7-13) is found primarily in cactus. It is likely that more betaxanthins will be isolated, but the compounds most important from a food colorant point of view are vulgaxanthin-1 and vulgaxanthin-2 from red beets (Fig. 7-12).

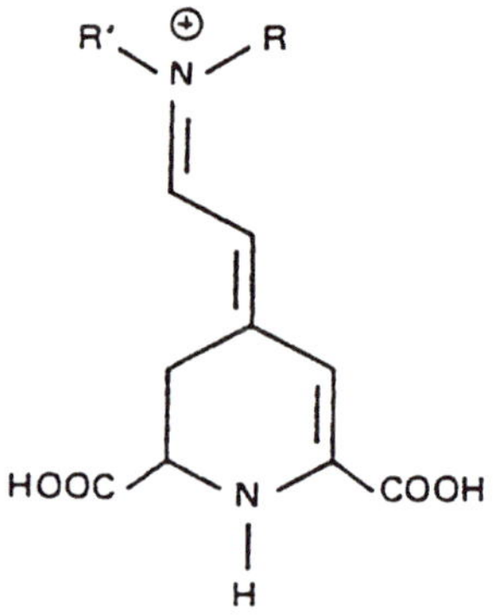

Fig. 7-9. The diazoheptamethin structure.

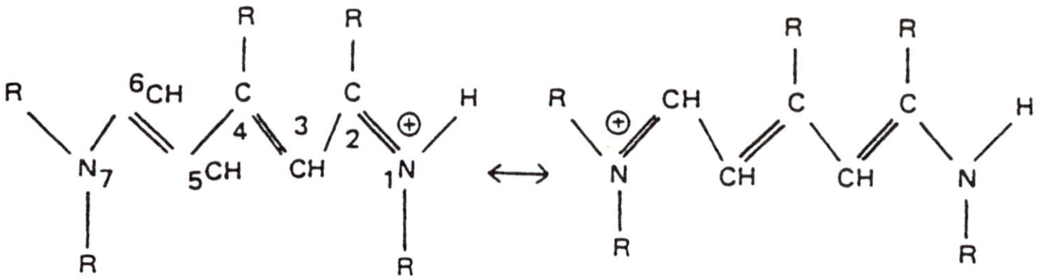

Fig. 7-10. Betacyanin resonating structures.

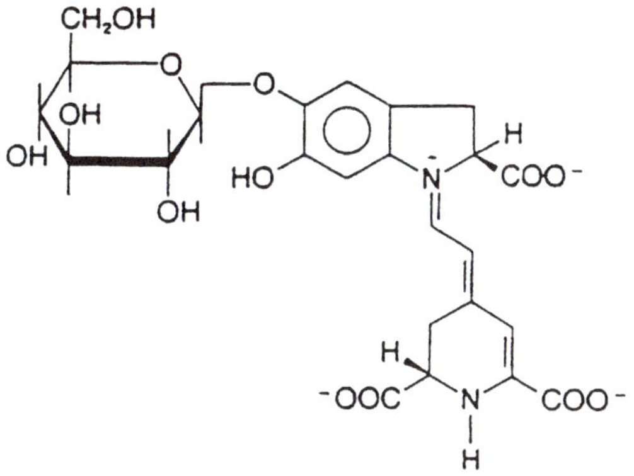

Fig. 7-11. The structure of betanin.

OCCURRENCE

The betalains are confined to 10 closely related families of the order *Caryophyllales*. The only foods containing betacyanins are the red beet, *B. vulgaris*; chard, *B. vulgaris*; cactus fruit, *Opuntia ficus-indica*; and pokeberries, *P. americana*. Pokeberries are not a normal food source, but the leaves are eaten as a green vegetable. Betalains are also found in a number of flowers and in the poisonous mushroom, *Amanita muscaria,* but again these are not normal food sources. The importance of betalains as a food colorant derives solely from extracts of red beets.

Red beets contain 75–95% betanin with the remainder being isobetanin, prebetanin, and isoprebetanin. The last two are the sulfate monoesters of betanin and isobetanin (Fig. 7-13), respectively. Beets generally contain both the red betacyanins and the yellow betaxanthins. Cultivars are available with different ratios of the red and yellow pigments; some cultivars contain only the yellow betaxanthins. Thus, it is possible by selecting appropriate cultivars of red beets and/or blending with extracts of yellow beets to provide a range of red to yellow colorants.

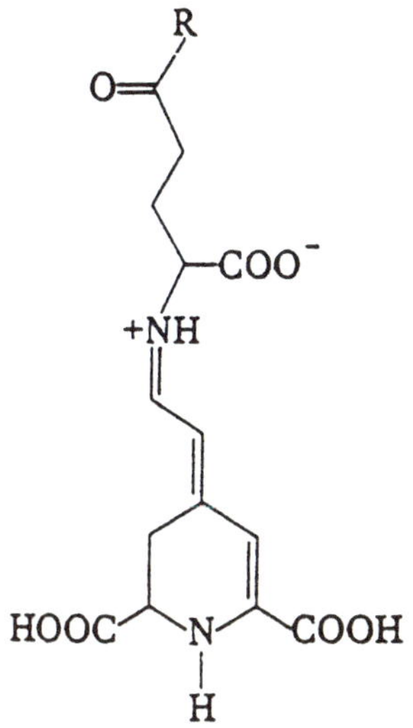

Fig. 7-12. Structures of vulgaxanthin-1 ($R = NH_2$) and vulgaxanthin-2 ($R = OH$).

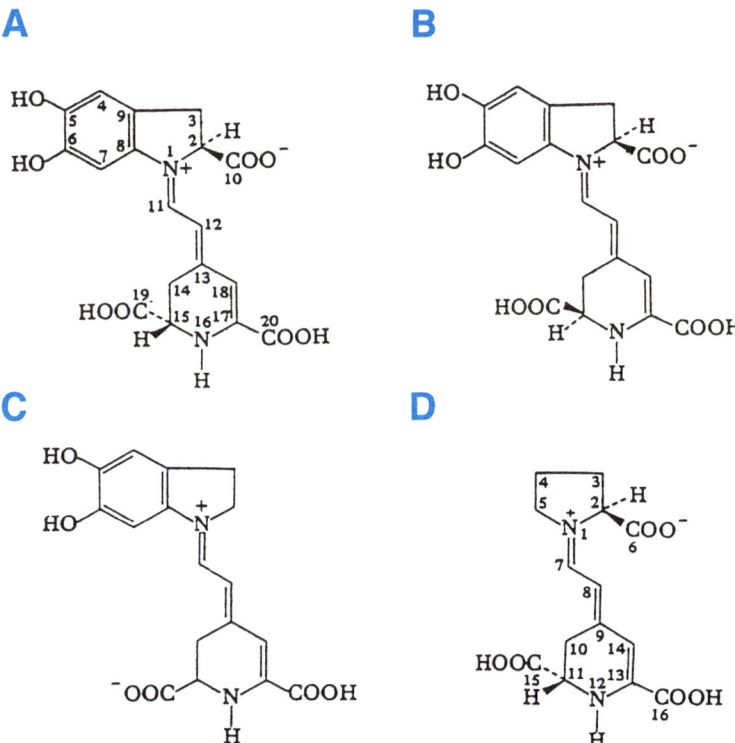

Fig. 7-13. Structures of betanidin (A), isobetanidin (B), 2-decarboxybetanidin (C), and indicaxanthin (D).

The anthocyanin/flavonoid and betacyanin/betaxanthin groups are mutually exclusive in the plant kingdom. Anthocyanins and betacyanins have never been reported in the same plant.

COMMERCIAL PREPARATION

Most countries allow two or three commercial color preparations from red beets. Liquid concentrates can be prepared by pressing blanched ground beets, filtering the liquid, and concentrating it under vacuum to 60–65% total solids. This product is classified by the FDA as a "vegetable juice" color additive as long as it meets the specifications outlined in 21 CFR 73.260. This vegetable juice is commonly spray dried with maltodextrin (a GRAS diluent) as a carrier and referred to as "beet powder." A second type of beet-based color is referred to as "dehydrated beets." Mature, sound-quality beets are dehydrated and ground. In the United States, this material must meet the FDA specifications outlined in 21 CFR 73.40. Dehydrated beet powder is generally used in dry mix products in which the brown notes and insolubles are not an issue—in chocolate-flavored beverage mixes, for example.

Beet juice contains considerable sugar, so a yeast fermentation is sometimes included to reduce the sugar content. The alcohol produced is mostly removed in the fermentation step. Beet juice usually exhibits a beet-like taste and odor due to the presence of *geosmin*. Fermentation with *Aspergilus niger*, *Candida utilis*, and *Saccharomyces oviformis* has been reported to produce product with improved stability and also to remove the geosmin. Traditionally, hydraulic or rotary presses have been used, but they recover only about 50% of the pigment. Recoveries as high as 90% can be accomplished by continuous diffusion processes. Several resin and other absorption processes such as Dowex 50 W followed by polyamide column chromatography with methanol as the eluent, polyacrylamide (Bio-Gel P-6), and numerous others (7) are available for purification of the pigments, but they are not usually permitted by law.

In vitro cell suspensions, reported to yield high levels of pigment production (19), have a number of advantages. The high growth rates of cultures make pigment production efficient, and the characteristic odor of geosmin is not produced. A number of pigments can be produced selectively as desired, but economic commercialization of cell culture production systems has yet to be demonstrated. However, there is a need for more highly purified and concentrated colorants.

Commercial beet powders usually contain 0.4–1.0% pigment expressed as betanin, 80% sugar, 8% ash, and 10% protein together with citric and/or ascorbic acid as a preservative. The tinctorial power is usually expressed as percent betanin. Betanin is identified as EU No. E 162 or CAS Reg. No. 7659-95-2.

Geosmin—A chemical found in red beets that contributes most of the beet flavor.

APPLICATIONS

Betanin preparations are water soluble with high tinctorial strength. They are relatively unchanged in color from pH 3 to 7 but are violet at pH values below 3 and bluer at pH values above 7. The absorption spectra of betanin at three pH values is shown in Figure 7-14. A shift in dominant wavelength toward the left indicates a yellower color and to the right a bluer color. The spectrum at pH 9.0 does show a small shift in color toward the bluish red.

Both the red and yellow pigments are thermolabile with and without the presence of oxygen and are also degraded by light. Metal cations such as iron, copper, tin, and aluminum accelerate the degradation. Within these limitations, beet preparations are ideally used to color products that have short shelf life; are packaged to reduce exposure to light, oxygen and high humidity; do not receive extended or high heat treatment; and are marketed in the dry state (7). Beet extracts have been suggested for dairy products such as ice cream and yogurt, salad dressings, frostings, cake mixes, gelatin desserts, meat substitutes, poultry meat sausages, gravy mixes, soft drinks, powdered drink mixes, marshmallow candies, hard candies, and fruit chews. An example of the use of tristimulus data to match colors was presented by von Elbe (20), who compared conventional products with those colored with betanin (Table 7-1). Betanin colorants can also be blended with other colorants to provide desired color matches.

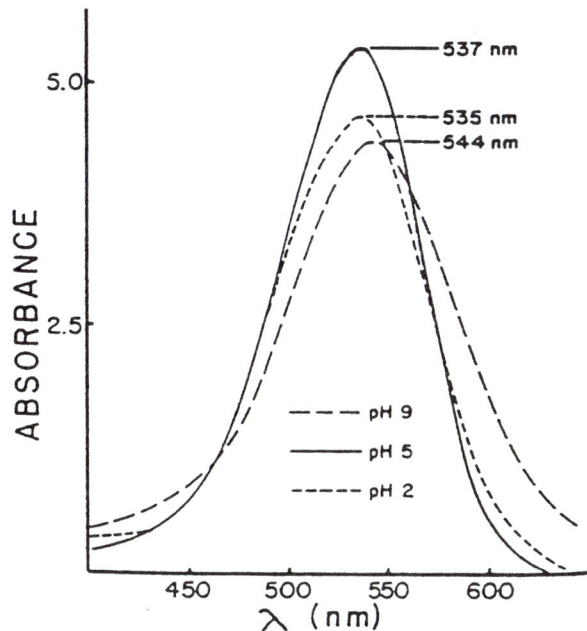

Fig. 7-14. Spectra of aqueous betanin solutions at pH 2.0, 5.0, and 9.0. Reprinted, by permisssion, from (20).

Table 7-1. Color Readings of Foods Colored with Betalains Compared with Conventional Samples[a]

Food	Color			
	L	a	b	θ
Bologna				
Conventional (150 ppm of SO2)	53.4	11.4	9.8	40.7
Betanin (33 ppm of SO2)	41.1	9.5	11.0	49.0
Sherbet, raspberry				
Conventional	53.3	33.0	−2.6	355.5
20 ppm betanin + 1.3 ppm bixin	57.5	33.0	−3.0	354.9
Gelatin dessert, raspberry				
Conventional	15.5	2.9	0.4	4.9
Betanin (48 ppm)	15.3	3.0	0.7	12.9

[a] From (20); used by permission.

References

1. Francis, F. J. 1987. Food colorants: Anthocyanins. Crit. Rev. Food Sci. Nutr. 28:273-314.
2. Francis, F. J. 1993. Polyphenols as natural food colorants. In: *Polyphenolic Phenomena*. A. Scalbert, Ed. Institut National de la Recherche Agronomique, Versailles, France.
3. Francis, F. J. 1986. *Handbook of Food Colorant Patents*. Food and Nutrition Press, Westport, CT.

4. Anonymous. 1995. *Concise Encyclopedia of Food and Nutrition.* A. H. Ensminger, M. E. Engsminger, J. E. Konlands, and J. A. K. Robson, Eds. CRC Press, Boca Raton, FL.
5. Harborne, J. B. 1993. *The Flavonoids: Advances Since 1980.* Chapman and Hall, New York.
6. Mazza, G. 1995. Anthocyanins in grapes and grape products. Crit. Rev. Food Sci. Nutr. 35:341-371.
7. Jackman, R. L., and Smith, J. L. 1992. Anthocyanins and betalains. In: *Natural Food Colorants.* G. A. F. Hendry and J. D. Houghton, Eds. Blackie Publishers, Glasgow, Scotland.
8. Bassa, Y., and Francis, F. J. 1987. Stability of anthocyanins from sweet potatoes in a model beverage. J. Food Sci. 52:753-754.
9. Shi, Z., Lin, M., and Francis, F. J. 1992. Stability of anthocyanins from *Tradescantia pallida.* J. Food Sci. 57:761-765.
10. Amic, D., Baranac, J., and Vukadinovic, V. 1990. Reactivity of some flavylium cations and corresponding anhydrobases. J. Agric. Food Chem. 38:936-940.
11. Bridle, P., Scott, K. G., and Timberlake, C. F. 1973. Anthocyanins in *Salix* species. New anthocyanin in *Salix purpurea* bark. Phytochemistry 12:1103-1106.
12. Cameira dos Santos, P. J., Brillouet, J. M., Cheyniere, V., and Moutounet, M. 1998. Detection and partial characterization of new anthocyanin-derived pigments in wine. J. Sci. Food Agric. (in press).
13. Sarni-Manchado, P., Fulcrand, H., Souquet, J. M., Chenier, V., and Moutounet, M. 1996. New wine anthocyanin-derived pigments: Comparison of their stability and color quality with that of grape anthocyanins. J. Food Sci. 61:938-941.
14. Jurd, L. 1972. Some advances in the chemistry of anthocyanin-type plant pigments. In: *The Chemistry of Plant Pigments.* C. O. Chichester, Ed. Academic Press, New York.
15. Kinsella, J. E., Kanner, J., Frankel, E., and German, B. Wine and health: The possible role of phenolics, flavonoids and other antioxidants. Presented at a symposium: Possible Health Effects of Components of Plant Foods and Beverages in the Diet. University of California, Davis, 1992.
16. Miniati, E., and Cole, R. 1993. Anthocyanins: Not only color for foods. In: *Proceedings of the First International Symposium on Natural Colorants.* F. J. Francis, Ed. The Hereld Organization, Hamden, CT.
17. Driver, M. W., and Francis, F. J. 1979. Stability of phytolaccanin, betanin, and FD & C Red No. 2 in dessert gels. J. Food Sci. 44:518-520.
18. Driver, M. W., and Francis, F. J. 1979. Purification of phytolaccanin by removal of phytolaccatoxin from *Phytolaccca americana.* J. Food Sci. 44:521-523.
19. Leathers, R. R., Davin, C., and Zryd, J. P. 1992. In vitro cell development. Biology 28:39-45.
20. von Elbe, J. H. 1975. Stability of betalains as food colors. Food Technol. 29(5):42-46.

CHAPTER 8

Chlorophylls, Haems, Phycobilins, and Anthraquinones

Chlorophylls

The chlorophylls are a group of naturally occurring pigments present in all photosynthetic plants, including algae and some bacteria. They are in greater abundance than any other organic pigment produced in nature. Hendry (1) estimated annual production at about 1,100,000,000 tons, as compared with carotenoids at 100,000,000 tons, with about 75% being produced in aquatic, primarily marine, environments. Obviously, as a source of raw material for food colorants, chlorophylls present no problem with supply.

CHEMISTRY

Five chlorophylls and five bacteriochlorophylls are known in nature. Only two, chlorophylls a and b (Fig. 8-1) are important as source material for food colorants. These come from land plants.

Chorophylls a and b have a complex structure, but they differ only by a $-CH_3$ and a $-CHO$ group on carbon 7. (The numbering system for chlorophylls is shown in Figure 8-2.) The chlorophyll a molecule is easily changed in the presence of acids to remove the magnesium ion, resulting in a derivative called pheophytin a. The phytyl group is easily removed to produce chlorophyllide a. If both magnesium and phytyl groups are removed, the compound is called pheophorbide a. A similar series of reactions occurs with chlorophyll b. The common names "chlorophyll," "chlorophyllide," "pheophytin," and "pheophorbide" were developed historically and have been retained even though the chemical nomenclature was changed by the *IUB-IUPAC Joint Commission on Nomenclature* in 1980. Hendry (1) provides a good description of the new nomenclature and the chemistry of chlorophylls.

The change to pheophytin is the major change that occurs when products containing chlorophyll are heated. The chlorophylls have a bright green color, whereas the pheophytins are a less-appealing olive green. Thus, the major efforts to maintain the attractive green color in vegetables during processing have involved trying to retain the magnesium in the molecule. Pretreatment and treatment during processing in alkaline solutions containing magnesium (the Blair Process

---

In This Chapter:

Chlorophylls
 Chemistry
 Preparation of Colorants
 Applications

Haems

Phycobilins
 Chemistry
 Extraction
 Applications

Anthraquinones
 Cochineal and Carmine
 Kermes
 Lac
 Alkannet

IUB-IUPAC—International Union of Biochemistry–International Union of Physics and Chemistry.

Chlorophyll a

Chlorophyll b

Phytyl group

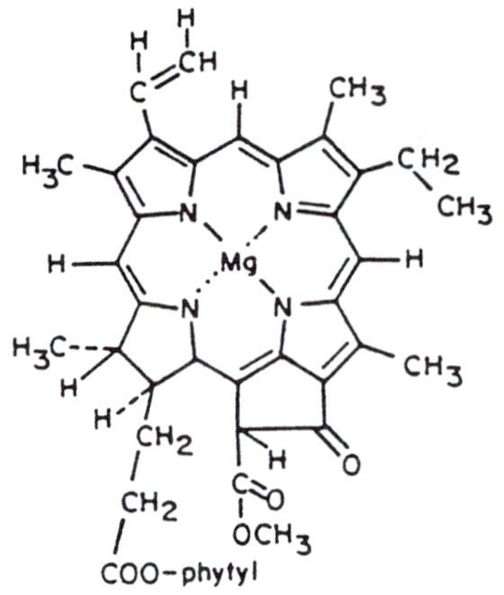

Fig. 8-1. Formulas for chlorophylls a and b and for phytol, from which the phytyl group is formed. Reprinted, by permission, from (2).

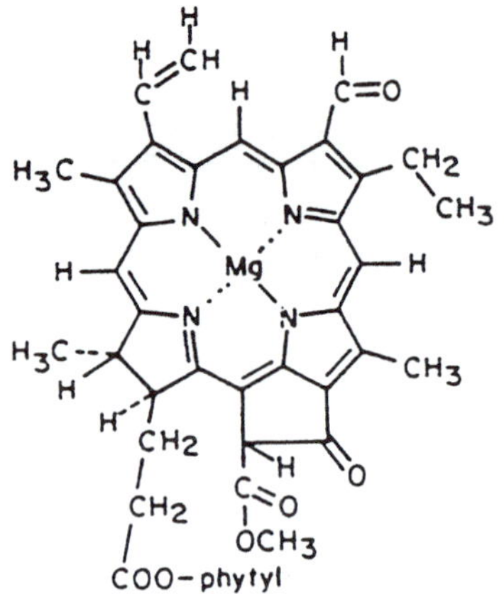

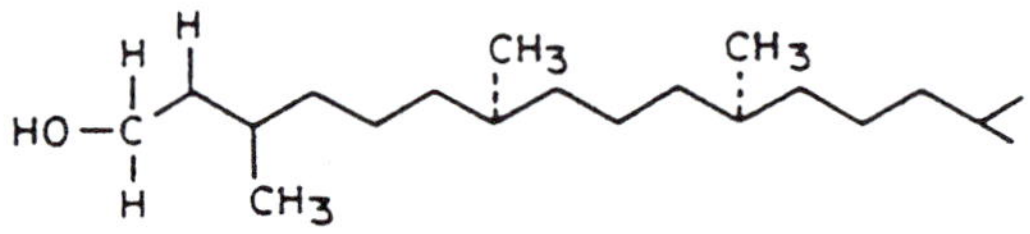

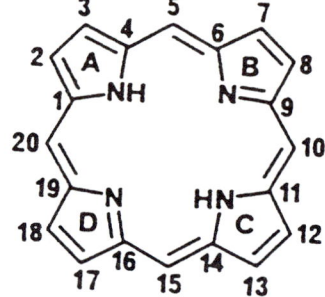

Fig. 8-2. Numbering system for the chlorophylls. Reprinted, by permission, from (1).

[3]) were successful for a short time after thermal processing. Enzymatic treatments to convert the chlorophyll to chlorophyllide (4) were attempted because the chlorophyllides have a color similar to that of chlorophylls and are more stable, and this did help to maintain the color. A combination of the two processes with high-temperature short-time processing (5) also maintained the color a little longer. But the retention of attractive color lasted only a few weeks, not long enough to provide an economic incentive, so the attempts were abandoned. More recent research (6–8) has involved attempts to replace the magnesium in the chlorophyll molecule with zinc or copper, since the green zinc and copper derivatives are known to be more stable. None of these processes has resulted in commercial success. The same problems with stability were experienced with attempts to make colorants containing chlorophyll itself.

PREPARATION OF COLORANTS

A third type of chlorophyll, chlorophyll c, occurs in the brown seaweeds that are used as a commercial source of alginates and also in single-celled phytoplankton. Although chlorophyll c is more stable than b or a, it has not been commercialized. Algae are already being grown as a source of carotenoids (Chapter 6) and conceivably could also be grown for chlorophyll content if the increased stability were sufficient to provide economic motivation. In a living cell, within the chloroplasts, the chlorophylls are complexed with but not covalently

bound to one of a series of polypeptides. The chlorophyll-polypeptide or chlorophyll-protein complexes are closely associated with carotenoids and tocopherols (vitamin E) and are actively involved in the processes of photosynthesis. Preparation of a colorant involves recovery and purification of an appropriate form of chlorophyll.

The plants currently being used to produce chlorophyll colorants include alfalfa (lucerne, *Medicago sativa*), nettles (*Urtica dioica*), and a series of pasture grasses such as fescue. Land plants are usually available for only a few weeks of the year, so the chlorophyll colorants are usually made from plants that are bulk-harvested and dried. The dried plants are ground and extracted with acetone or a chlorinated hydrocarbon solvent, followed by washing and solvent removal. A yield of 20% of a mixture of chlorophylls, pheophytins, and other degraded chlorophyll compounds is usually obtained; this is further processed into an oil-soluble or water-soluble colorant. The dry residue can be purified by treatment with a water-immiscible solvent to obtain an oil-soluble preparation known as metal-free pheophytin or can be acidified in the presence of copper salts to produce an oil-soluble copper pheophytin. The dry residue can also be saponified to replace the phytyl group with sodium or potassium to form water-soluble compounds. After further purification, the colorant is gray-green and may be marketed as water-soluble and metal-free. Acidification in the presence of copper salts converts the pigments to more stable green pigments known as copper chlorophyllins. It is not commercially practical to produce a colorant containing chlorophyll itself because of the instability of the magnesium compound.

APPLICATIONS

The copper-substituted derivatives of pheophytin are relatively stable to light and mineral acids and are more stable than the metal-free pheophytins and pheophorbides. The major portion of the estimated $15,000,000 (in 1994 dollars) in annual sales of colorants derived from chlorophyll is from the water-soluble forms (1). The major portion of both oil-soluble and water-soluble forms is used in dairy products, edible oils, soups, chewing gum, sugar confections, and drinks. A minor proportion is used in cosmetics and toiletries.

Chlorophyll colorants are widely used in Europe and other parts of the world. In the United States, they are not permitted for foods. They are permitted in dentifrices but must be from alfalfa.

Haems

The term *haem* or *heme* is used to describe iron derivatives of the cyclic tetrapyrrole protoporphin with the structure shown in Figure 8-3. The numbering system is similar to that for chlorophyll (Fig. 8-2). This structure is basic to a number of compounds such as chlorophyll, hemoglobin, and cytochrome (which are important in photosynthesis, energy transport, and oxygen transport, respectively), and several

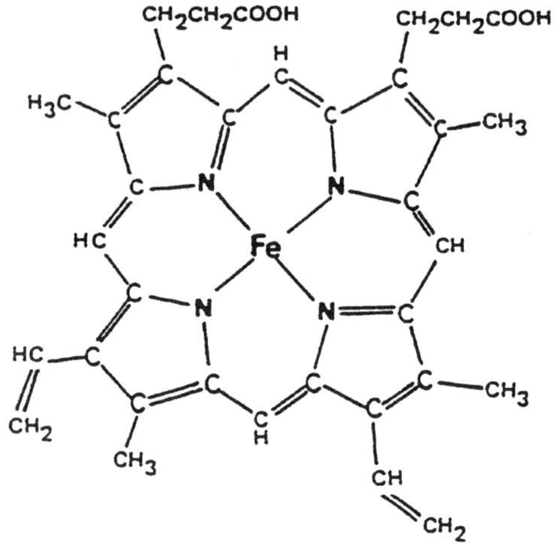

Fig. 8-3. Structure of haem.

enzymes. Hemoglobin, probably the best known, is composed of a haem group associated with a protein. Hemoglobin forms a loose association with oxygen as the basis for oxygen transport and carbon dioxide removal and is remarkably stable *in vivo* but not *in vitro*. This introduces some difficulty in using haems as colorants.

Preparation of haem from blood has been known for centuries. It is accomplished by the addition of whole blood to glacial acetic acid saturated with salt at 100°C. Crystalline haem precipitates out, but it is not stable enough to function as a food colorant. More stable compounds can be produced by allowing other ligands such as carbon monoxide, nitrous oxide, hydroxides, and cyanide to displace the oxygen from the central iron atom. Several patents exist (9), falling into three broad areas: 1) treatment of whole blood with a variety of reagents, 2) stabilization of the color with a variety of ligands, and 3) purification of haem derivatives. Treatment of whole blood to remove the hemoglobin has considerable appeal because the iron is in a nutritionally available form, useful as a dietary supplement, and the protein has a high biological value.

It would be still another advantage if the preparation could be made stable enough to act as a food colorant. A number of the patents claim that the preparations are suitable for coloring products that normally contain meat, meat analogues, sausages, etc. The recovery of the haem pigments can be as simple as direct spray-drying of whole blood. The haem products can be concentrated by simple dehydration using alcohols. Several patents describe the production of adducts with carbon monoxide, nitrous oxide, and hydroxy and carboxy compounds, but none of these processes seems to have been commercialized.

The use of whole blood as a food ingredient is a very old custom. The Europeans are famous for "blood pudding," which is black in color because hemoglobin is denatured in the cooking process. Houghton (10) reported an interesting application of naturally occurring haem pigments, namely, the bile pigments bilirubin and biliverdin. They occur in gall stones and hair balls and are in demand in Chinese medicine as aphrodisiacs.

Blood is permitted as a food ingredient if it has been collected and processed in an appropriate sanitary manner, but colorants derived from blood are not permitted in the United States.

Phycobilins

Phycobilins are major biochemical components of blue-green, red, and cryptomonad algae. They are colored, fluorescent, water-soluble pigment-protein complexes. Phycobilins can be classified into three major groups according to color: phycoerythrins are red with a bright orange fluorescence; phycocyanins and allophycocyanins are both

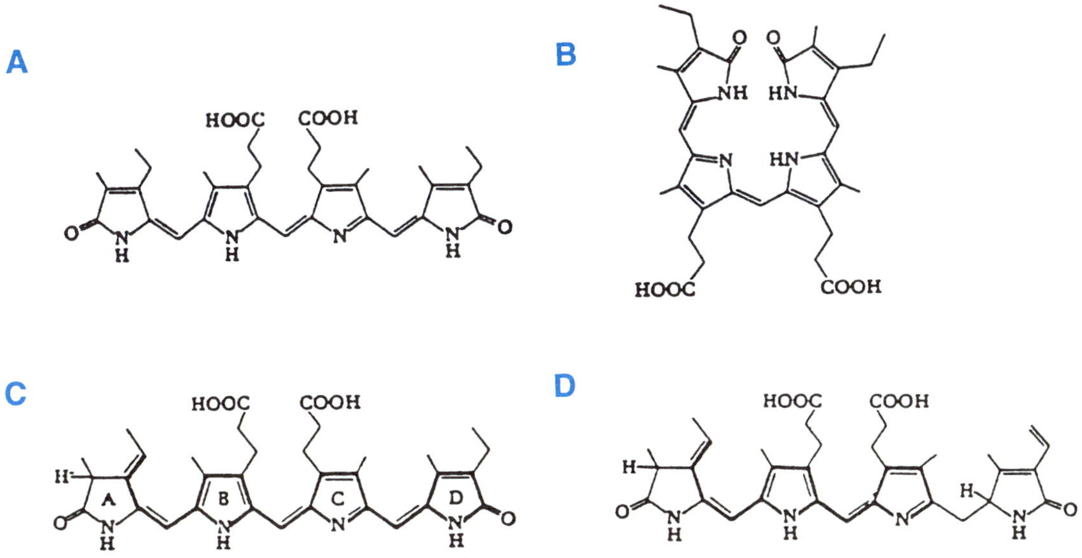

Fig. 8-4. Structure of the phycobilins. A, a bilin in the conventional linear form. B, the same structure in a cyclic form to show its relationship with the porphyrins. C, a blue phycocyanobilin pigment. D, a red phycoerythrobilin pigment.

blue and fluoresce red. The range of color and the apparent stability of the pigments make this group attractive as food colorants.

CHEMISTRY

The structures for the two *chromophores* of the phycobilins, termed phycocyanobilin and phycoerythrobilin, are shown in Figure 8-4C and D. Phycocyanins and allophycocyanins share the same chromophore; the differences in color are due to the different protein groups. Figure 8-4 shows a bilin written in the conventional linear form (A) and also in a cyclic, possibly truer, form (B), showing its relationship to other pigments such as chlorophyll and hemoglobin. The attachment of the bilin chromophore to its protein is very stable because there are covalent bonds between the ethylidene group on position 3 (Fig. 8-2) of the bilin and a cysteine group in the protein. A second covalent bond may be in position 18. The major function of the bilin pigments is to act as light absorbers in the energy transport system.

EXTRACTION

Phycobilin preparations can be obtained by simply freeze-drying an algal cell suspension, which produces a highly colored powder. The chlorophyll and carotenoid components can be removed by centrifuging after breaking the algal cells. The centrifugate is a brightly colored solution of water-soluble protein, of which 40% may be phycobilin. Further purification by precipitation with ammonium sulfate followed by ion-exchange yields almost pure phycobilin. The free chromophore

Chromophore—A general term for the portion of a molecule that contributes most of the color.

can be obtained by extended refluxing in methanol. An alternative to extensive purification would be to grow an organism such as the alga *Cyanidium caldarium*, which, on addition of 5-aminolevulinic acid to the growth medium, yields pure phycocyanobilin (10).

Another approach is to use the existing technology for growth of the alga *Spirulina platensis* in open ponds. This organism has been used for many years in Africa and Mexico, probably because of its high protein content and digestibility. It is sold in health food stores as a dietary supplement. The organism is very salt tolerant, which helps to prevent contamination with other microorganisms. The cells are simply harvested and dried. Recent research has shown that it is possible to extract phycobilins from *Spirulina*, and this is currently being commercialized.

Another unicellular organism, the red alga *Porphyridium cruentum*, can also be grown in open ponds or in tubular reactors for production of phycobilin and phycoerythrin. A number of patents exist for the extraction, stabilization, purification, and use of phycocyanins from *Spirulina* and *Aphanotheca nidulans*, and three firms are marketing such products (9).

APPLICATIONS

The presence of a protein in the pigment suggests that the colorant would be used for products requiring a minimum of heat treatment. But the fact that it takes 16 hr in boiling methanol to separate the protein from the chromophore suggests that applications could include products with mild heat treatment. Suggested applications for phycocyanin colorants include chewing gums, frozen confections, soft drinks, dairy products, sweets, and ice cream. Although the use of phycoerythrin has not been fully explored, it seems like a good candidate for a red colorant.

Phycobilins have several industrial applications, such as fluorescent tracers in biochemical research, fluorescence-activated cell sorting, and fluorescence microscopy (10). Development of other markets will certainly help to create the critical volume necessary for production of food colorants. Colorants containing phycobilins are currently not permitted in the United States.

Anthraquinones

The anthraquinones and naphthoquinones have been the subject of considerable research interest lately. Twelve patents were issued in the 1963–1984 era (9), and a large number of new compounds in this class have been reported (11). Interest in them as food colorants stems mainly from their stability and high tinctorial strength. They were the chromophores of choice in the "linked polymer" concept of colorants pioneered by the Dynapol Company, although this concept is no longer being commercialized (9).

COCHINEAL AND CARMINE

Cochineal extract (CI Natural Red 4, CI No. 75470, EEC No. E 120) is a red solution obtained by treating the dried bodies of female cochineal insects, particularly *Dactylopius coccus* Costa, with aqueous ethanol. After removal of the alcohol, the preparation, called cochineal extract, contains approximately 2–4% carminic acid (Fig. 8-5) as the main colorant compound. It takes 50,000–70,000 insects to produce 1 lb of the colorant.

Cochineal extract is a very old colorant. References to it are found as far back as 5000 BC, when Egyptian women used it to color their lips. It was introduced into Europe by Cortez after he found it in Mexico in 1518. The Aztecs had been using it for many years, and the

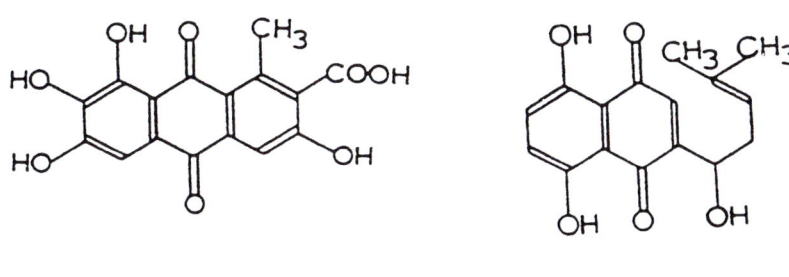

Fig. 8-5. Structures of some anthraquinones.

native Mexicans cultivated the cochineal insect on the aerial parts of cactus. The Spaniards guarded the secret of cochineal, and by 1700 as much as 500,000 lb of cochineal (probably the dried bodies of the insects rather than the colorant) were being shipped to Spain from Mexico each year. Other areas, including the East and West Indies, Central and South America, Palestine, India, Persia, Europe, and Africa, later developed the ability to produce cochineal. The cochineal trade peaked in 1870 and then declined rapidly due to the introduction of synthetic colorants in 1856 and their subsequent growth. Today, Peru is the major supplier of the dried insects, with an annual production of about 400 tons, which constitutes 85% of the world production. Mexico and the Canary Islands share the remaining 15%.

American cochineal is the most significant form today. It is obtained from insects of the family Dactylopiidae, primarily *Dactylopius coccus* Costa, found as parasites on the aerial parts of cactus, *Opuntia* and *Nopalea* species, espcially *N. coccinelliferna*. There are other sources. Similar pigments, produced by insects from the families *Coccoidea* and *Aphidoidea*, have various historical names. Armenian Red is obtained from the insect *Porphyrophera hameli*, which grows on the roots and stems of several grasses in Ajerbaizan and Armenia. Polish cochineal is obtained from the insects *Margaroides polonicus* or *Porphyrophera pomonica*, found on the roots of *Scleranthus perennis*, a grass growing in central and eastern Europe.

Cochineal and its derivatives are staging a comeback today as food colorants because of their superior technological properties (such as stability, clarity, and desirable hue) and the influence of the "natural" trend.

Chemistry. Cochineal is extracted from the bodies of female insects just before egg-laying, at which time the insects may contain as much as 22% of their dry weight as pigment. One may wonder at the biological significance of this; presumably, it is a protection from predators. Historically, the insects were extracted with hot water, and the colorants were known as "simple" extracts of cochineal. More recently, purification methods including proteinase enzyme treatments have produced colorants sometimes known as the "carmines of cochineal." The word "carmine" has been used as a general term for this class of anthraquinones, but the more usual meaning of the term "carmine" (CAS Reg. No. 1390-65-4) in the United States denotes a calcium or calcium-aluminum lake of carminic acid. In some other countries, a magnesium lake of carminic acid may be called "carmine." The lakes usually contain about 50% carminic acid (Fig. 8-5), which is the pigment in cochineal. In the United States, the FDA requires a minimum of 50% carminic acid.

Solutions of carminic acid at pH 4 show a range of color from pale yellow to orange, depending upon what concentration is used. They complex with metals to produce stable brilliant red hues. Complexes with tin and aluminum produce the most desirable hues, and nearly all commercial preparations contain aluminum. A range of hues from "strawberry" to "black currant" can be produced by adjusting the ratio of carminic acid to aluminum. The color is essentially independent of the pH value, being red at pH 4 and bluish red at pH 10. Carminic acid is usually available as an aqueous solution with a colorant content of about 2–5%. It may also be spray-dried. Formulations may also contain propylene glycol, glycerine, citric acid, and sodium citrate.

Carmine is very stable to heat and light, resistant to oxidation, and not affected by sulfur dioxide. The presence of other metal ions may shift the color slightly toward the blue. The colorant is usually supplied as an alkaline solution with a carminic acid content of 2–7%. The content of carminic acid is usually the way of specifying the carmine's strength, but Schul (12) suggested that the price of carmine should be based on its color strength, not on the carminic acid content, and provided details of a method. The FDA-approved procedures are specified in 21 CFR Section 73.100. Traditionally, ammonia was used as an alkalizing agent, but recently formulations with potassium hydroxide, spray-dried with maltodextrin as a carrier, have become available. They may also contain sodium hydroxide, ammonium hydroxide, and glycerol. Carmine lakes are more cost-effective as colorants than carminic acid because of the added costs to separate and concentrate carminic acid.

Applications. Carmine is considered to be technologically a very good food colorant. Its nonkosher status is an important limitation, although some kosher carmines exist commercially today. Carmine-based products are ideally suited for food systems with a pH above 3.5, including comminuted meat products such as sausages, processed poultry products, surimi, and red marinades. Other important uses are in jams and preserves, gelatin desserts, baked goods, confections, icings, toppings, dairy products, and uncarbonated drinks and related products. The level of usage varies with the product and is usually 0.05–1.0%. Cochineal and carmine are permitted as food colorants in the United States.

KERMES

Kermes is a well-known red colorant in eastern Europe. It is derived from the insects *Kermes ilicis* or *Kermococcus vermilis*, which are found in the aboveground portions of several species of oak, particularly *Quercus coccifera*, the Kermes oak. The pigment in kermes is kermesic acid, the aglycone of carminic acid (Fig. 8-5). It also occurs as an isomer, ceroalbolinic acid (Fig. 8-5). The pigment is obviously closely related to carminic acid, and its properties are very similar. Colorants from kermes are not permitted in the United States.

LAC

Lac is a red colorant obtained from the insect *Laccifera lacca* found on the trees *Schleichera oleosa*, *Ziziphus mauritiana*, and *Butea monosperma*, which grow in India and Malaysia. The lac insects are better

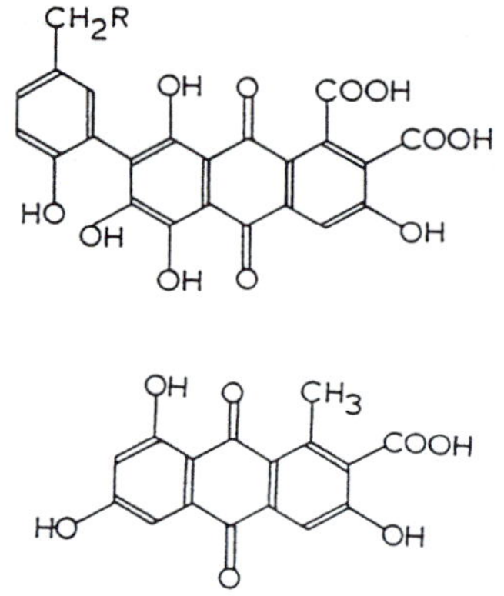

Fig. 8-6. Structures of the anthraquinone pigments in lac. The top structure represents laccaic acids A–C, and the bottom one represents laccaic acid E. For laccaic acid A, R = $CH_2NHCOCH_3$; for laccaic acid B, R = CH_2OH; for laccaic acid C, R = $CHNH_2COOH$; for laccaic acid E, R = CH_2NH_2.

known as a source of shellac. Lac pigments are a complex mixture of anthraquinones (Fig. 8-6) and also contain the closely related pigments erythrolaccin and deoxyerythrolaccin (Fig. 8-5). The chemistry and applications of lac are similar to those of cochineal. Colorants from lac are not permitted in the United States.

ALKANNET

Alkannet is a red pigment (Fig. 8-5) extracted from the roots of *Alkanna tinctoria* Taush and *Alchusa tinctoria* Lom from southern Europe. The pigment is almost insoluble in water but readily soluble in organic solvents. It has been used in Europe to color confectionery, ice cream, and wines, but it is not permitted in the United States.

References

1. Hendry, G. A. F. 1996. Chlorophylls and chlorophyll derivatives. In: *Natural Food Colorants*. G. A. F. Hendry and J. D. Houghton, Eds. Blackie Publishers, Glasgow, Scotland.
2. Aronoff, S. 1966. In: *The Chlorophylls*. L. P. Vernon and G. P. Seely, Eds. Academic Press, New York.
3. Blair, J. S. 1940. Color stabilization of green vegetables. U.S. patents 2,186,003 and 2,189,774.
4. Clydesdale, F. M., and Francis, F. J. 1968. Chlorophyll changes in thermally processed spinach as influenced by enzyme conversion and pH adjustment. Food Technol. 22(6):135-138.
5. Clydesdale, F. M., Fleishman, D. L., and Francis, F. J. 1968. Maintenance of color in processed green vegetables. Food Prod. Dev. 4:126-138.
6. Laborde, L. F., and von Elbe, J. H. 1994. Effect of solutes on zinc complex formation in heated green vegetables. J. Agric. Food Chem. 42:1096-1099.
7. Laborde, L. F., and von Elbe, J. H. 1994. Chorophyll degradation and zinc complex formation with chlorophyll derivatives in heated green vegetables. J. Agric. Food Chem. 42:1100-1103.
8. Laborde, L. F., and von Elbe, J. H. 1996. Method of improving the color of containerized green vegetables. U.S. patent 5,482,727.
9. Francis, F. J. 1986. *Handbook of Food Colorant Patents*. Food and Nutrition Press, Westport, CT.
10. Houghton, J. D. 1996. Haems and bilins. In: *Natural Food Colorants*. G. A. F. Hendry and J. D. Houghton, Eds. Blackie Publishers, Glasgow, Scotland.
11. Francis, F. J. 1996. Lesser known food colorants. In: *Natural Food Colorants*. G. A. F. Hendry and J. D. Houghton, Eds. Blackie Publishers, Glasgow, Scotland.
12. Schul, I. J. 1993. An ancient but still young colorant. In: *Proceedings of the First International Symposium on Natural Colorants*. F. J. Francis, Ed. The Hereld Organization, Hamden, CT.

CHAPTER 9

Turmeric, Carthamin, and *Monascus*

Turmeric

Turmeric (CI Natural Yellow 3, CI No. 75300, EEC No. E 100), also called "curcuma," is the dried ground rhizomes of several species of *Curcuma longa*, a perennial herb of the Zingiberaceae family. It is native to northern Asia and is cultivated in China, India, South America, and the East Indies. Annual production is about 165,000 tons.

In This Chapter:
Turmeric
  Properties
  Applications
Carthamin
  Properties and Application
Monascus
  Chemistry
  Commercial Production
  Applications

PROPERTIES

Turmeric is a bright yellow powder with a characteristic odor and taste. There are three major pigments: curcumin (CAS Reg. No. 458-37-7), demethoxycurcumin, and bisdemethoxycurcumin (Fig. 9-1). These pigments, together with the flavor compounds turmerone, cineol, zingeroni, and phellandrene, constitute turmeric (the ground rhyzomes) and turmeric oleoresin. These two forms are listed in the CFR in Sections 73.600 and 73.615, respectively. The oleoresin is prepared by extraction of dried rhizomes with one or a combination of solvents, including methanol, ethanol, isopropanol, ethyl acetate, hexane, methylene chloride, ethylene dichloride, and trichloroethylene. Other methods include extraction with ether, evaporation of the ether, and suspension of the residue in vegetable oil. Powdered turmeric preparations, consisting of the powdered rhizome or the oleoresin standardized with maltodextrin, usually contain 8–9% curcumin. Liquid concentrates of oleoresin suspended in ethanol and/or propylene glycol with a polysorbate emulsifier usually contain 0.5–25% curcumin. A variety of other carriers such as edible oils and fats and monoglycerides are also available (1). The FDA-approved extraction procedures are specified in CFR Section 73.615.

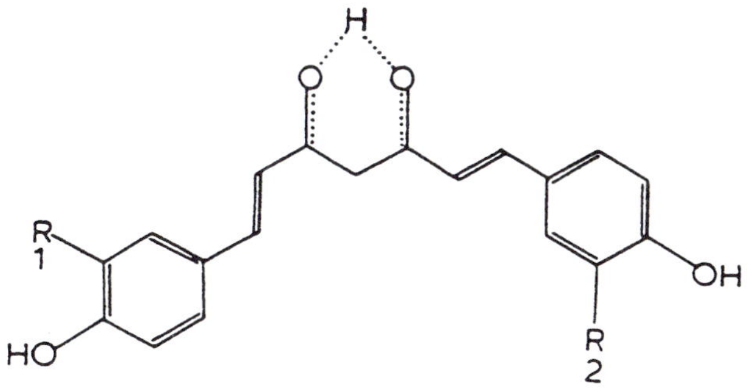

APPLICATIONS

Turmeric and turmeric oleoresins are both unstable in light and in alkaline conditions; thus, several patents involve formulation with citric, gentisic, or gallic acid and tannic acid polyphosphate for stabilization. Sodium citrate, waxy maize starch, and an emulsifier are

Fig. 9-1. Pigments of turmeric. Curcumin: R1 = R2 = OCH3. Demethoxycurcumin: R1 = H, R2 = OCH3. Bisdemethoxycurcumin: R1 = R2 = H.

sometimes added. Both powder and liquids are susceptible to degradation by oxidation. Both show good tinctorial strength, with turmeric usually used in the 0.2–60 ppm range and the oleoresin in the 2–640 ppm range. The tinctorial strength is usually expressed as percent curcumin, even though preparations from different geographical regions differ in the relative content of curcumin and demethoxycurcumin. Both turmeric and the oleoresin produce bright yellow to greenish yellow shades and are sometimes used as a replacement for FD&C Yellow No. 5. Curcumin is insoluble in water, but water-soluble complexes can be made by complexing with heavy metals such as stannous chloride and zinc chloride to produce an intense orange colorant. These compounds, however, are not permitted by the FDA. Curcumin colorants can also be prepared by absorbing the pigments on finely divided cellulose. Water-soluble forms of curcumin can be obtained that contain alcohols, propylene glycol, and Polysorbate 80.

The major applications of turmeric are to color cauliflower and pickles and as an ingredient in mustard. It is also used alone or in combination with other colorants such as annatto in spices, ice cream and yogurt, cheeses, baked goods, confectionery, cooking oils, and salad dressings.

Carthamin

Carthamin, also called "carthemone," "carthemus," "carthamic acid," or "safflor red" is identified as CI Natural Red 26, CI 75140. It is a yellow to red preparation from safflower flowers, *Carthemus tinctorius*, of the family Compositae, cultivated extensively in Europe and America. Its use as a textile colorant dates back to antiquity under a variety of names such as Spanish saffron, African saffron, American saffron, thistle saffron, false saffron, bastard saffron, and Dyer's saffron.

PROPERTIES AND APPLICATION

Carthamin is the only member of the chalcone group of the flavonoid pigments that has been suggested as a food colorant. It contains three chalcones, the red carthamin, safflor yellow A, and safflor yellow B (Fig. 9-2). Fresh yellow flower petals contain precarthamin, which oxidizes to form the red carthamin. Under acid conditions, carthamin equilibrates to two isomers, red carthamin and yellow isocarthamin. The earlier patents for carthamin involved simple aqueous extraction of the petals or crushing the petals before extraction to allow for oxidation. The three main pigments can be purified by passage through a styrene resin bed. Purification and stabilization can be greatly enhanced by absorption of the pigments on cellulose powder. Apparently, cellulose has a great affinity for carthamin; this is known as the "Saito effect." According to Saito and Fukushima (2), "the effect is so strong that the carthamin may be re-

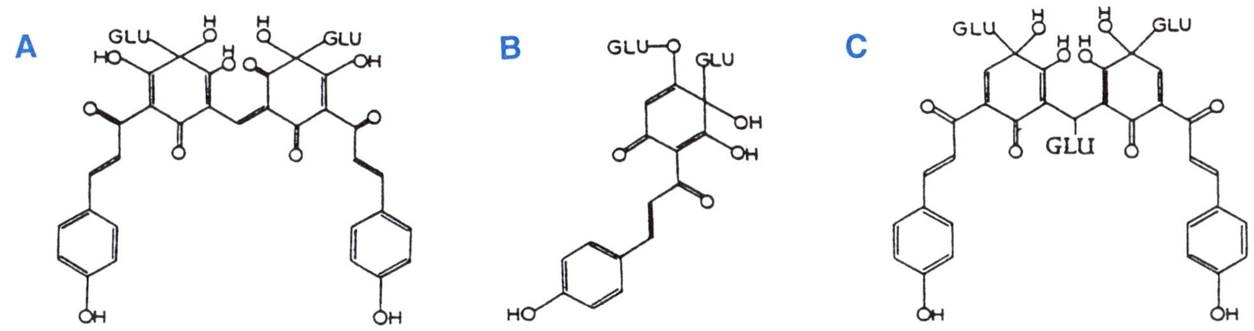

Fig. 9-2. Pigments of carthamin. A = Carthamin; B = safflor yellow A; C = safflor yellow B.

tained for more than a thousand years without appreciable change to the coloration." However, no storage data were provided to substantiate this claim. The cellulose absorption method was adapted to large-scale production using a methanolic extract and a cellulose column in acid media (3). Another method using a cellulose derivative, diethylaminoethylcellulose, yielded pure precarthamin, carthamin, safflor A, and safflor B. Tissue culture approaches have been successful in producing the safflower pigments as well as some novel closely related pigments (1).

Carthamin has been suggested as a colorant for pineapple juice, yogurt, butter, liqueurs, confectionery, etc. The yellow to red range of colors gives it some flexibility. Currently, it is not approved for use in foods in the United States.

Monascus

The genus *Monascus* includes several fungal species that grow on a number of carbohydrate substrates, especially steamed rice. It is well known in the southern and eastern Asian countries and was mentioned as far back as 1590 as a Chinese medicine (4). Traditionally, *Monascus* species were grown on rice, and the whole mass was eaten as an attractive red food. It could also be dried, ground, and incorporated into other foods.

CHEMISTRY

Monascus species produce six pigments with the structures shown in Figure 9-3. Monascin

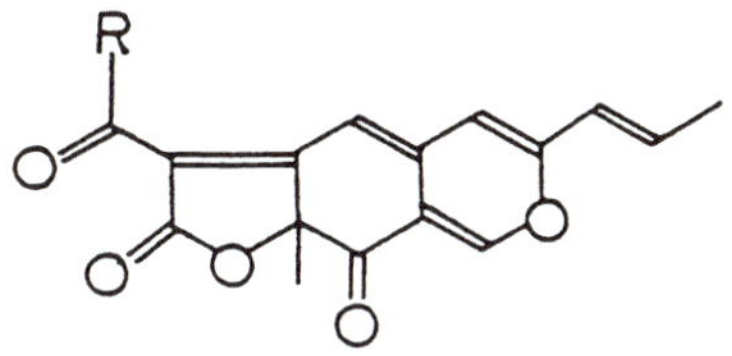

Monascin, R = C_5H_{11}
Ankaflavin, R = C_7H_{15}

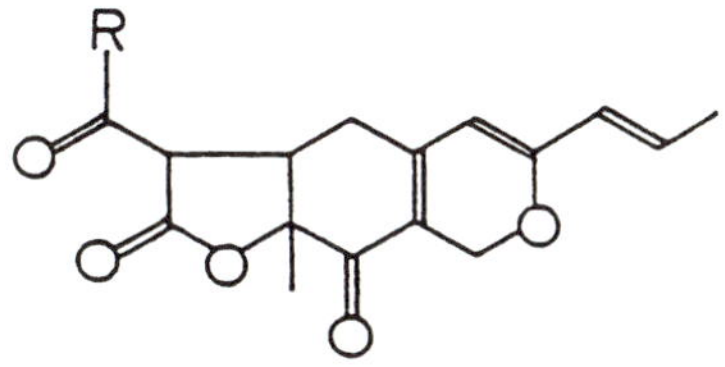

Rubropunctatin, R = C_5H_{11}
Monascorubin, R = C_7H_{15}

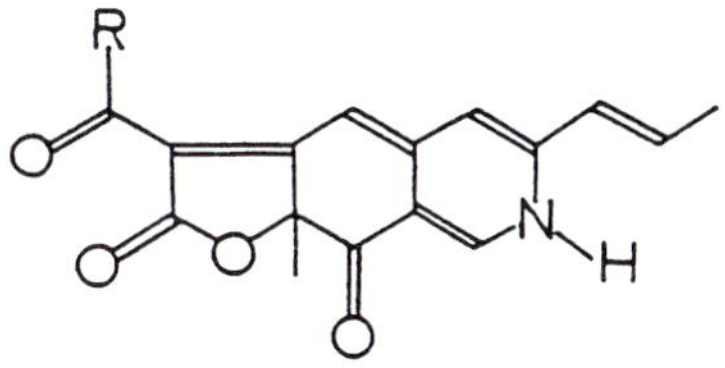

Rubropunctamine, R = C_5H_{11}
Monascorubramine, R = C_7H_{15}

Fig. 9-3. Pigments of *Monascus*.

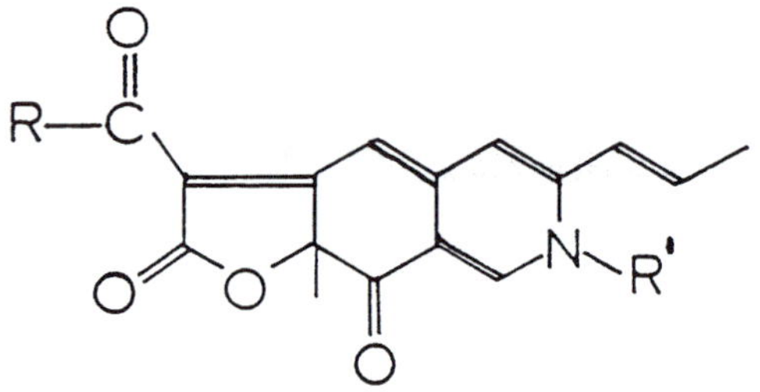

Fig. 9-4. A red derivative of the monascin pigment. R represents an aliphatic radical and R´ represents a compound of the formula HN-R, i.e., an amino sugar, a polymer of an amino sugar, or an amino alcohol. From (5).

and ankaflavin are yellow; rubropunctatin and monascorubin are red; and rubropunctamine and monascorubramine are purple. The red and yellow pigments are considered to be normal secondary metabolites of fungal growth, and the purple pigments may be formed by enzymatic modification of the red and yellow pigments. The red pigments are very reactive and readily react with amino groups via a ring opening and a Schiff rearrangement to form water-soluble compounds such as that in Figure 9-4. The Monascus pigments have been reacted with amino groups, polyamino acids, amino alcohols, chitin amines or hexamines, proteins, peptides and amino acids, sugar amino acids, browning reaction products, aminoacetic acid, and amino benzoic acid (1). Obviously, a number of derivatives are possible, and they have been shown to have greater water solubility, thermostability, and photostability than the parent compounds.

COMMERCIAL PRODUCTION

Monascus colorants are currently being produced in Japan, China, and Taiwan. Traditionally, *Monascus* has been grown on solid cereal substrates such as rice and wheat and the entire mass used as a food or a food ingredient. However, it became obvious that the fungus could be grown in liquid or fluid solid-state media and optimized for pigment production. Much research has been devoted to the determination of the optimal conditions for pigment production on a wide variety of cereal substrates. Production of either the red or yellow pigments can be maximized to produce a range of red to yellow colors. For example, Tadao et al (6) reported that *M. anka* grown on bread produced only yellow pigments. Early production methods were unsatisfactory because the fungus produced an antibiotic, which is considered an undesirable food ingredient. Recent strain selections and control of growth conditions have eliminated the antibiotic problem. Han and Mudgett (7) concluded that *M. purpureus* (ATCC 16365) was the most appropriate organism for solid-state fermentation and confirmed that it had no antibiotic activity. The fungus also produces alcohol, which in smaller quantities increases pigment production and in larger concentrations decreases it. *Monascus* produces a range of compounds normal to growth metabolism such as enzymes, coenzymes, *monascolins* (which modify fat metabolism), antihypertensive agents, and flocculants. These compounds optimize growth, but they cause strain selection problems and must be removed from products used for food colorants.

Monascolins—A group of microbial metabolites produced in association with the production of *Monascus* pigments.

APPLICATIONS

The *Monascus* pigments are readily soluble in ethanol and only slightly soluble in water. Ethanolic solutions are orange at pH 3.0–4.0, red at 5.0–6.0, and purplish red at 7.0–9.0. The pigments fade under prolonged exposure to light, but in 70% alcohol they are more stable than in water. The pigments are very stable to temperatures up to 100°C in neutral or alkaline conditions.

Monascus colorants offer considerable advantages since they can be produced in any quantity on inexpensive substrates. Their range of colors from yellow to red and their stability in neutral media is a real advantage. Sweeny et al (8) suggested *Monascus* colorants for processed meats, marine products, jam, ice cream, and tomato ketchup. They should be appropriate for alcoholic beverages such as saki and also for koji, soy sauce, and kamboko. Considerable interest has been shown in the *Monascus* group; 38 patents were issued in the years 1969–1985 (9). Currently, *Monascus* colorants are not permitted in the United States.

References

1. Francis, F. J. 1992. Miscellaneous colorants. Chap. 7 in: *Natural Food Colorants*. G. A. F. Hendry and J. D. Houghton, Eds. Blackie & Sons, Ltd., Bishopbriggs, Glasgow, Scotland.
2. Saito, K., and Fukushima, A. 1988. On the mechanism of the stable red colors of cellulose-bound carthamin. Food Chem. 29:161-175.
3. Saito, K., and Fukushima, A. 1986. Effect of external conditions on the stability of enzymatically synthesized carthamin. Acta Soc. Bot. Pol. 55:639-651.
4. Wong, H. C., and Koehler, P. E. 1981. Mutants for monascus-pigment production. J. Food Sci. 46:956-957.
5. Moll, H. R., and Farr, D. R. 1976. Red pigment and process. U.S. patent 3,993,789.
6. Tadao, H., Shima, T., Suzuki, T., Tsukioka, M., and Takahashi, T. 1981. Edible yellow pigment from Monascus. Japanese patent 81,006,263.
7. Han, O. H., and Mudgett, R. E. 1992. Effects of oxygen and carbon dioxide on Monascus growth and pigment production in solid state fermentation. Biotechnol. Prog. 8:5-12.
8. Sweeny, J. G., Estrado-Valdez, M. C., Iacobucci, G. A., Sato, H., and Sakamura, S. J. 1981. Photoprotection of the red pigments of *Monascus ank* in aqueous media by 1,4,6-trihydroxynaphthalene. Agric. Food Chem. 29:1189-1193.
9. Francis, F. J. 1986 *Handbook of Food Colorant Patents*. Food and Nutrition Press, Westport, CT.

Caramel, Brown Polyphenols, and Iridoids

Caramel

Caramel (CI Natural Brown 10, EEC No. E 150) is a brown colorant obtained by heating sugars. The official FDA definition lists the sugars:

> The color additive caramel is the dark brown liquid or solid resulting from carefully controlled heat treatment of the following food-grade carbohydrates: dextrose, invert sugar, lactose, malt syrup, molasses, starch hydrolysates and fractions thereof, or sucrose.

The heating of sugar preparations to create brown, flavorful, and pleasant-smelling products has been practiced in home cooking for centuries. The sauces or candies are known as caramels. Commercial practices to prepare caramel colorants began in Europe about 1850. The first caramel colorants were prepared by heating sugars in an open pan, but in view of their popularity, modifications were soon introduced.

In This Chapter:

Caramel
  Commercial Preparation
  Chemistry
  Applications

Brown Polyphenols
  Cacao
  Tea

Iridoid Pigments
  Chemistry
  Production
  Applications

COMMERCIAL PREPARATION

An appropriate amount of carbohydrate is added to a reaction vessel at 50°C; the temperature is raised to 100°C; and the reactants are added. After appropriate heating, the mass is cooled and filtered. The pH and specific gravity are adjusted with acids, alkalis, or water to meet customer specifications. The carbohydrate hydrolysates may be obtained from sucrose, corn, wheat, and tapioca. High glucose content is desirable because caramelization occurs only through the monosaccharide. Caramels can be prepared by using reactants such as carbonates, hydroxides, ammonium compounds, sulfites, and both ammonium and sulfite compounds. The type and concentration of the reactant determine the properties of the caramel colorant. For instance, a direct relationship exists between reactant concentration and color intensity. Too little reactant results in an underreacted product with a lighter color and lower viscosity. Too much reactant produces a product that is very dark and very viscous or even solid.

CHEMISTRY

In 1980, the Joint Expert Committee on Food Additives (JECFA) of the Food and Agriculture Organization/World Health Organization recommended that further information on the chemical properties of caramel be obtained in order to establish a suitable classification and specification system. The International Technical Caramel Committee undertook an extensive research program to provide this information. This resulted in the grouping of the caramel formulations into four classes, depending on the net ionic charge and the presence of reactants (Table 10-1).

TABLE 10-1. Classes of Caramel Formulations

Class	Charge	Reactants	Use
I	−	No ammonium or sulfite compounds	Distilled spirits, desserts, spice blends
II	−	Sulfite compounds	Liqueurs
III	+	Ammonium compounds	Baked goods, beer, gravies
IV	+	Both sulfite and ammonium compounds	Soft drinks, pet foods, soups

Heating carbohydrates in the presence of a reactant produces a wide range of chemical compounds. Complete characterization is not feasible; thus, a series of profiles were developed using 157 samples from 11 manufacturers in seven countries. Since Caramel Color IV accounts for 70% of all caramel colors manufactured, it was chosen as a prototype for this ambitious undertaking. The profiles developed were screening by high-performance liquid chromatography; size fractionation by ultrafiltration to provide low, intermediate, and high molecular weight groupings; and subfractionation by cellulose chromatography. Each fraction was examined by a series of sophisticated analytical approaches. These data, combined with 11 physical characteristics, confirmed that the four groupings provided real and reproducible classifications (1).

Color is a physical parameter of interest to formulators of food products. It could, of course, be specified by any of the conventional tristimulus color scales, but the color can be accurately specified by a simpler method (2). The color of caramel results from a large number of chemical chromophores; thus, different formulations do not show markedly different spectral curves and do not change shape with concentration. Curves of different concentrations can be superimposed on each other by plotting log absorbance against concentration. This is the "optical signature" of a colorant. When log absorbance is plotted against wavelength for caramel solutions, the result is a straight line that moves up or down depending on the concentration. The slope of the line changes according to the hue of the solution. This makes it possible to determine two colorimetric indices, the hue index and the tinctorial power, from two simple absorbance measurements.

The *hue index* is $10 \log(A_{0.51}/A_{0.61})$, where 0.51 = absorbance at 0.51 μm and 0.61 = absorbance at 0.61 μm.

The *tinctorial power*, K, is $K_{0.56} = (A_{0.56}/cb)$, where 0.56 = absorbance at 0.565 μm, c = concentration (g/L), and b = cell thickness (cm).

Both the hue index and the tinctorial power should reflect the visual appearance over the whole visual spectrum.

Hue index—A value describing the yellow to red hue of a colorant formulation.

Tinctorial power—A value describing the ability of a colorant to color a product.

One of the major considerations in the research initiated by JECFA was the safety aspect of the caramel colorants. The program resulted in the publication of 11 papers in the same 1992 (Vol. 30) issue of the journal *Food Chemical Toxicology*; seven of them were on toxicology. Caramel was given a clean bill of health, and the JECFA assigned an ADI of 0–200 mg/kg per day. Some safety issues affecting caramel and other colorants are discussed in Box 10-1.

APPLICATIONS

Caramel colorants must be compatible with the food products in which they are used. Compatibility consists essentially of the absence of flocculation and precipitation in the food. These undesirable effects result from charged macromolecular components of caramel interacting colloidally with the food. Hence, the net ionic charge of the caramel color macromolecules at the pH of the product in which the caramel is used is a prime determinant of compatibility (5).

Caramel color is freely soluble in water but insoluble in most organic solvents. In concentrated form, the colorant has a characteristic burnt taste that is not detectable at the levels normally used in

---

Box 10-1. How Pure Are Color Additives?

The lack of classification is common to a number of natural extracts and also to some manufactured colorants. For example, no manufacturer has attempted to provide a chemical profile of the components of extracted colorants from grapes. This is understandable in view of the complexity of the profile (Chapter 7).

A somewhat similar situation occurs with a synthesized product such as FD&C Yellow No 5. When it was approved for food use in 1916, the manufacturers wanted to produce as pure a product as possible. The impurities they had to cope with were unreacted ingredients and three compounds produced by side reactions. In the usual process of chemical synthesis, the desired reaction does not normally go to 100% completion with no side reactions, and it is the job of the manufacturer to remove the unwanted by-products. But the increasing sophistication of analytical procedures that can routinely detect one part per billion, or even as low as one molecule (3), has allowed researchers to detect more impurities. Today, FD&C Yellow No. 5 has been shown to have as many as 17 components present, some in exceedingly low concentrations (4). Other FD&C colorants are probably similar. The same situation exists with synthetic β-carotene.

Modern analytical instrumentation has made it possible to identify the impurities in chemicals produced with rigid process control, as illustrated with FD&C Yellow No 5 and β-carotene. For caramel and colorants produced from grapes, it is not yet possible to specify in detail the structures of all the compounds present.

As we gain more knowledge of impurities, the question arises: What do impurities mean from a safety point of view? For one thing, if the studies are to mean anything, the chemical profile of a given additive in routine manufacture must be the same as that used for the toxicology studies. It is economically infeasible to remove all impurities in food additives. Thus, some form of risk assessment is unavoidable. Fortunately, all four products mentioned above have been given a clean bill of health.

foods and beverages. Commercial preparations vary from 50 to 70% total solids and have a range of pH values, depending on the type of caramel. Over 80% of the caramel produced in the United States is used to color soft drinks, particularly colas and root beers.

Brown Polyphenols

CACAO

The cacao plant, *Theobroma cacao,* has been suggested as a potential source of brown colorants. The cacao pods, beans, shells, cotyledons, husks, and stems all contain a complex mixture of polyphenols such as cyanidin glycosides, leucoanthocyanins, (–)-epicatechins, quercitin glycosides, and the acyl acids *p*-coumaric and gentisic. Colorants prepared from cacao are likely to contain a complex mixture of leucoanthocyanins, flavonoid polymers, and catechin-type polymers.

The suggested extraction methods involve hot acid, alkaline water, or ethanol followed by filtration and concentration. A more reddish extract can be obtained by prior roasting of the beans, followed by extraction with hot aqueous alkaline solutions. The colorant is suitable for most foods for which a brown color is desired. Cacao products are permitted in the United States as food ingredients but not as a source of colorants.

TEA

Extracts of tea, *Thea sinensis,* have been used as brown colorants for centuries. The polyphenols in tea comprise a very complex mixture, including the glucosides and rhamnoglucosides of myricetin, quercetin, and kaempferol; di-*C*-glycosylapigenins; 7-*O*-glucosylisovitexin; epicatechin; epigallocatechins; epicatechin gallate; gallic acid; chlorogenic acid; ellagic acid; coumarylquinic acid; and many related compounds. In black tea, they may act as precursors to the poorly defined pigments thearubin and theaflavin. Black tea also contains theaflavin gallate, digalloylbisepigallocatechin, triacetinidin, flavanotropolone, and flavanotropolone gallate.

Preparation of extracts of tea is usually a simple extraction of the leaves and stems with warm water or ethanol, followed by filtration and concentration. Extracts of tea are legal food ingredients in the United States, but specific colorants prepared from tea products are not. This may appear to be a fine distinction, but it depends on whether the colorant is classified as an ingredient or an additive.

Iridoid Pigments

CHEMISTRY

Colorants from saffron have enjoyed good technological success and gustatory appeal, but their high price has led to searches for other sources of the same pigments. The pigments, but not the flavor, can be obtained in much larger quantities from the fruits of the gardenia, or Cape jasmine, plant. The fruits of gardenia, *Gardenia jasminoides,* contain three major groups of pigments: crocins (Chap. 6), iridoids, and flavonoids. The carotenoids, crocins, and related compounds develop during weeks 8–23 of growth; the iridoid pigments develop one to six weeks after flowering. Structures of nine iridoid pigments, a series of flavonoid compounds, are shown in Figure 10-1. The structure of five flavonoids isolated from *Gardenia fosbergii* is shown in Figure 10-2. This is a different species from *G. jasminoides,* but the flavonoid compounds in closely related species tend to be similar. The flavonoids contribute a pale yellow color and the carotenoids an orange color.

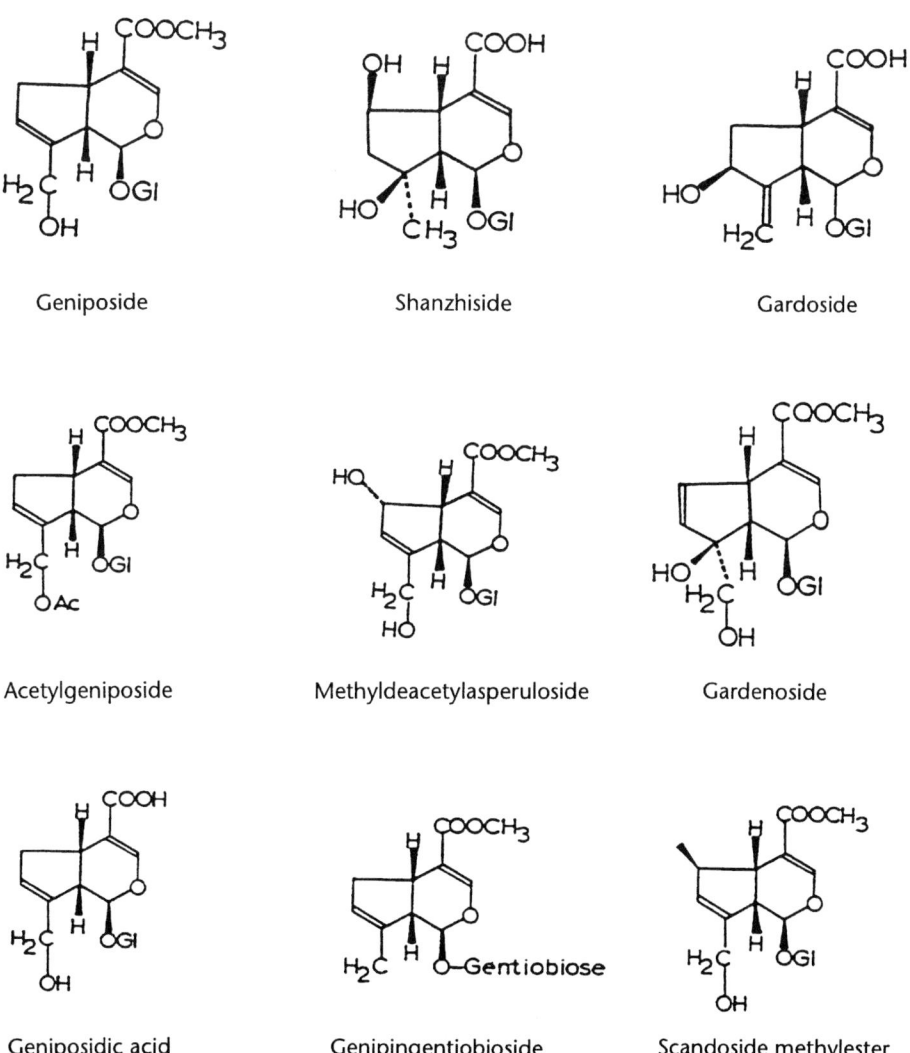

Fig. 10-1. Structures of nine iridoid pigments in *Gardenia jasminoides.* Gl = glucose.

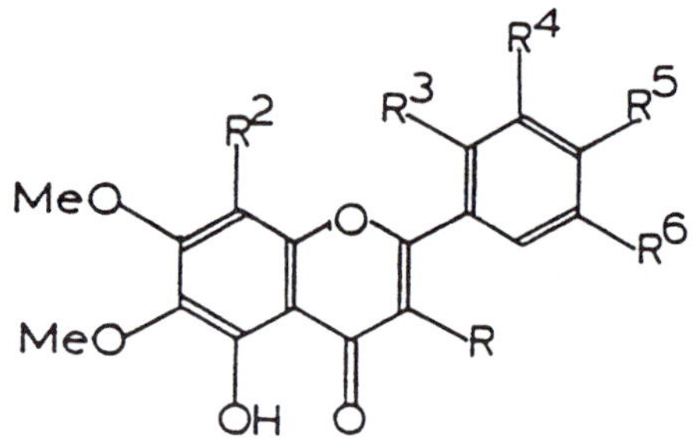

Compound	R¹	R²	R³	R⁴	R⁵	R⁶
1	H	H	H	OMe	OMe	OMe
2	OMe	H	H	OH	OMe	OH
3	H	H	OMe	OMe	H	OH
4	OMe	H	H	OMe	OMe	OH
5	OMe	OMe	H	H	OH	H

Fig. 10-2. Structures of five flavonoid pigments in *Gardenia fosbergii*.

PRODUCTION

The iridoid pigments are interesting as food colorants because a range of colors from green to yellow, red, and blue can be produced (6). Patent literature suggests extraction of the fruit with water, treatment with enzymes having β-glycosidic or proteolytic (bromelain) activity, followed by reaction with primary amines from amino acids or a protein such as soy. Manipulation of the reaction conditions, such as time, pH, temperature, oxygen content, degree of polymerization, and conjugation of the primary amino groups, allows a range of colorants to be produced. Several patents involve culturing extracts of gardenia with microorganisms such as *Bacillus subtilis, Aspergillus japonicus,* or a species of *Rhizopus*. Other patents involve hydrolysis of the iridoid glycoside geniposide by the action of β-glucosidase to produce genipin, which can be reacted with taurine to produce a blue colorant. Other amino acids such as glycine, alanine, leucine, phenylalanine, and tyrosine can be reacted to produce brilliant blue colorants. They are claimed to be stable for two weeks at 40°C in 40% ethanol. Four greens, two blues, and one red have been commercialized in Japan.

APPLICATIONS

Colorant preparations from gardenia have been suggested for use with candies, sweets, colored ices, noodles, imitation crab, fish eggs, glazed chestnuts, beans, dried fish substitutes, liqueurs, baked goods, etc. Because of their wide range of colors and their apparent stability, colorants from gardenia appear to have good potential. They are not approved for use in the United States.

References

1. Licht, B. H., Orr, J., and Myers, D. V. 1992. Characterization of caramel color, IV. Food Chem. Toxicol. 30:365-373.
2. Linner, R. T. 1971. Caramel coloring: A new method of determining its color hue and tinctorial power. Am. Soft Drink J. 1971 (5):26-30.
3. Francis, F. J. 1996. Safety of food colorants. In: *Natural Food Colorants*, 2nd ed. G. A. F. Hendry and J. D. Houghton, Eds. Blackie Publishers, Glasgow, Scotland.
4. Kassner, J. E. 1987. Modern technologies in the manufacture of certified food colors. Food Technol. 41(4):74-76.
5. Myers, D. V., and Howell, J. C. 1992. Characterization and specifications of caramel colours: An overview. Food Chem. Toxicol. 30:359-363.
6. Francis, F. J. 1996. Less common food colorants. In: *Natural Food Colorants*, 2nd ed. G. A. F. Hendry and J. D. Houghton, Eds. Blackie Publishers, Glasgow, Scotland.

CHAPTER 11

Miscellaneous Colorants

Inorganic Colorants

Several of the colorants described in this section, such as titanium dioxide and carbon black, are obviously colorants, but others may be used primarily for other purposes. For instance, zinc oxide and calcium carbonate may be added to foods as nutritional supplements and calcium carbonate as a pH adjuster; thus, their contribution to color is minimal. Still others, such as talc, are indirect additives used in processing or packaging. They do contribute incidentally to appearance, but it is not their primary function. There is a long list of compounds in this category (listed in 21 CFR 178.3297), as described in Chapter 4.

TITANIUM DIOXIDE

Titanium dioxide (CI Pigment White, CI No. 77891, CAS Reg. No. 1 3463-67-7, EEC No. E 171, Titanic Earth) is a large industrial commodity. Its current world production is estimated at nearly 4 million tons, but only a small percentage is used as a food colorant. Commercial TiO_2 is obtained from the naturally occurring mineral ilmenite ($FeTiO_2$) and occurs in three crystalline forms (anatase, brookite, and rutile), with anatase as the form most common in industry. The only form that currently meets FDA specifications is synthetically produced anatase. In addition, it is a soft particle, whereas rutile is much harder and coarse.

TiO_2 is the whitest pigment known today, with a hiding power five times greater than its closest rival, zinc oxide. This is the reason for its importance in paint formulations. TiO_2 is a very stable compound with excellent stability toward light, oxidation, pH changes, and microbiological attack. It is virtually insoluble in all common solvents, but it does dissolve slowly in hydrofluoric acid and hot, concentrated sulfuric acid. Water-dispersible and oil-dispersible forms are available, usually in water, glycerol, propylene glycol, sugar syrups, vegetable oils, or polyglycerol esters of fatty acids. Xanthan gum may be added as a stabilizer and potassium sorbate, citric acid, and methyl and ethyl parabens as preservatives. TiO_2 is used in confectionery, baked goods, cheeses, icings, and numerous pharmaceuticals and cosmetics. The specifications for TiO_2 for use in foods (Table 11-1) are similar to those of many other food additives.

In This Chapter:

Inorganic Colorants
  Titanium Dioxide
  Carbon Black
  Ultramarine Blue
  Iron Oxides
  Talc
  Zinc Oxide
  Calcium Carbonate
  Silver
  Silicon Dioxide
  Others

Organic Colorants
  Fruit and Vegetable Extracts
  Riboflavin
  Corn Endosperm Oil
  Algae Products
  Cottonseed Products
  Shellac
  Octopus Ink and Squid Ink

TiO₂ is allowed in foods up to 1% of the weight of the final product. It is often used in combination with FD&C lakes in tabletted products because the covering power of the mixture is greater than that of either the pure colorant or the lake alone. TiO₂ may also be used alone or in sugar syrups as a subcoating for tabletted products. It has been subjected to a number of toxicological studies and has been found to be nontoxic (1).

CARBON BLACK

Carbon Black is a large-volume industrial commodity, but its food usage is very small. Food-grade Carbon Black is derived from vegetable material, usually peat, by complete combustion to residual carbon. The powder colorant has a very small particle size, usually less than 5 µm, and consequently is very difficult to handle. Therefore, it is usually sold to the food industry in the form of a viscous paste, with the colorant suspended in glucose syrup. Little safety data are available. In the 1970s, when the GRAS list was being reviewed, toxicological data were requested in view of the theoretical possibility of contamination with heterocyclic amines. Apparently, the cost of obtaining the data was higher than the entire annual sales of food grade Carbon Black, so the tests were never done. Carbon Black is not currently permitted in the United States.

Carbon Black is a very stable and technologically a very effective colorant. It is widely used in Europe and other countries in sugar confectionery.

TABLE 11-1. Specifications for Titanium Dioxide Used in Foods

Component	Amount
Lead (as Pb)	10 ppm
Arsenic (as As)	1 ppm
Antimony (as Sb)	2 ppm
Mercury (as Hg)	1 ppm
Loss on ignition at 800°C, after drying for 3 hr at 105°C	0.5%
Water-soluble substances	0.5%
Acid-soluble substances	0.5%
TiO₂ content, after drying for 3 hr at 105°C	99.0%

ULTRAMARINE BLUE

Ultramarine Blue (CI No. 77007, E 180) is a synthetic blue pigment of rather indefinite composition. The ultramarine pigments are aluminosulfosilicates with empirical formulas approximated as $Na_2Al_6Si_6O_{24}S_3$. They are primarily crystalline in structure and related to the zeolite compounds. Ultramarine Blue is well known as a cosmetic and is intended to resemble the colorants produced from the naturally occurring semiprecious gem lapis lazuli.

The basic ingredients for the production of the ultramarines are kaolin (China clay), silica, sulfur, soda ash, and sodium sulfate, plus a reducing agent such as rosin or charcoal pitch. The ingredients, temperature, time, cooling rate, and processing variables determine the color. The basic product of high firing temperatures (up to 800°C) is Ultramarine Green (CI 77013), which can be transformed into Ultramarine Blue (CI 77007). With further treatment, Ultramarine Violet and Ultramarine Red can be produced. The colors are believed to result from the resonant polysulfide linkages. All four are important as cosmetic colorants, but the only food use is in salt intended for animal use at the 0.5% w/w level.

IRON OXIDES

The iron oxides represent a group of synthetic colorants. Although they exist in nature, difficulties in purification make the naturally occurring forms unacceptable. The iron oxides are known under a variety of names, such as CI Pigment Black 11 and CI Pigment Browns 6 and 7 (CI No. 77499), CI Pigment Yellows 42 and 43 (CI No. 77492), and CI Pigment Reds 101 and 102 (CI No. 77491). The chemical composition varies with the method of manufacture but can be represented by the empirical formula $FeO \cdot xH_2O$, $Fe_2O_3 \cdot xH_2O$, or some combination. Most are produced from ferrous sulfate $FeSO_4 \cdot 7H_2O$, and the most common forms are yellow hydrated oxides (ochre) and the brown, red, and black oxides.

Iron oxides are very stable compounds, insoluble in most solvents but usually soluble in acids. Their main uses are in cosmetics and drugs, but they are allowed (CI Nos. 77491, 77492, 77499) in dog and cat foods at levels up to 0.25% by weight of the finished food.

TALC

Talc (CI Pigment White, CI No. 77018, CAS Reg. No. 14807-96-6) is a naturally occurring magnesium silicate, $3MgO \cdot 4SiO_2H_2O$, sometimes containing a small amount of aluminum silicate. It is a large industrial commodity produced in many countries, particularly France, Italy, India, and the United States. The lumps are known as soapstone or steatite and the fine powders as talc, tateum, or French chalk. The best grades of talc are white crystalline powders of particle size 74 µm or less. Its major uses are as a dusting powder in medicine; as a white filler in paints, varnishes, and rubber; and as a lubricant in molds for manufacturing. It is used as a release agent in the pharmaceutical and baking industries as well as in coatings for rice grains. Its composition is shown in Table 11-2.

TABLE 11-2. Typical Composition of USP Talc[a]

Component	Amount
Silicon dioxide (SiO_2), %	60.13
Magnesium oxide (MgO), %	32.14
Calcium oxide (CaO), %	0.399
Aluminum oxide (Al_2O_3), %	1.84
Ferric oxide (Fe_2O_3), %	0.15
Acid solubles, %	<2.0
Water solubles, %	<0.1
Loss on ignition, %	4.9
Lead, ppm	<5
Arsenic, ppm	<1

[a] From (2).

ZINC OXIDE

Zinc oxide (CI Pigment White, CI No. 77947, CAS Reg. No. 1314-13-2) is a white or yellowish white, odorless, amorphous powder. It is insoluble in water but soluble in most mineral acids and alkali hydroxides. Zinc oxide is the most important white powder used in the cosmetic industry. It has the advantages of brightness, ability to provide opacity without blue undertones, and antiseptic and healing effects. It does not have the hiding power of titanium dioxide but sometimes is used as a whitener in wrappers for food. Zinc oxide is also added as a nutritional dietary supplement.

CALCIUM CARBONATE

Calcium carbonate (CI Pigment White 18, CI No. 77220, EEC No. E 170, CAS Reg. No. 471-34-1) is a fine white powder prepared by precipitating $CaCO_3$ from solution in three industrial processes: 1) as a by-product in the lime soda process, 2) precipitation from $CaCl_2$ in the calcium chloride process, and 3) precipitation from $CaOH_2$ in the carbonation process. Detailed specifications are listed in the *Food Chemicals Codex* (3). Calcium carbonate occurs naturally as limestone and marble, but the impurities make it unacceptable as a food ingredient. As a whitener, calcium carbonate has a minor role in foods. Its major roles are as a pH adjuster, nutritional dietary supplement (primarily for cereals), dough conditioner, firming agent, and yeast food. There are no limitations on its use other than Good Manufacturing Practices. Detailed specifications are listed in the *Food Chemicals Codex*. It is not currently listed as a food color additive by FDA and cannot be used specifically for that purpose in foods. However, it is permitted in drugs and cosmetics under 21 CFR 73.1070.

SILVER

Silver (EEC No. E 174) is a fine crystalline powder prepared by the reaction of silver nitrate with ferrous sulfate in the presence of nitric, phosphoric, or sulfuric acid. Polyvinyl alcohol is sometimes added to prevent the agglomeration of crystals and the formation of amorphous silver. Its primary use is as a fingernail polish colorant. Colloidal silver is sometimes used as a bactericidal aid in purifying solutions. In 1995, the FDA (21 CFR 176.170) approved the use of silver chloride-coated titanium dioxide as a preservative in polymer latex emulsions used in the coating of food-contact papers. It is to be used at a level not exceeding 2.2 ppm, based on the silver ion concentration in the dry coating.

Silver dragees are small silver-coated candy balls used as an ornament and ingredient in confectionery and baked goods in Europe and other areas of the world. Silver in foods is not permitted by FDA.

SILICON DIOXIDE

Silicon dioxide is an amorphous material produced synthetically by either a vapor-phase hydrolysis process yielding fumed (colloidal) silica or by a wet process that yields precipitated silica, silica gel, or hydrous silica. Fumed silica is produced as an anhydrous material, whereas the wet-process products are obtained as hydrates or contain water absorbed on the surface. Fumed silica occurs as a hygroscopic, white, fluffy, nongritty powder of very fine particle size. The wet-process silicas occur as hygroscopic, white, fluffy powders or white microcellular beads or granules. All of these forms of silicon dioxide are insoluble in water and in organic solvents but are soluble in hydrofluoric acid and in hot concentrated solutions of alkalis.

The colorant uses of silicon dioxide are minimal, but it is used widely in foods as an anticaking agent, defoaming agent, conditioning agent, and chillproofing agent in malt beverages.

OTHERS

Alumina. Alumina (EEC No. E 173) consists essentially of aluminum hydroxide $Al_2O_3 \cdot xH_2O$. It is a white, odorless, tasteless powder.

Aluminum powder. Aluminum powder (CI pigment Metal 1, CI No. 77000, EEC No. E 173) consists of finely powdered aluminum. It is permitted in the United States for external use but not in foods.

Gold. Gold (powder) is not listed for use in foods in the United States. In some countries, gold may be allowed in some alcoholic beverages, usually in the form of flakes.

Organic Colorants

FRUIT AND VEGETABLE EXTRACTS

Fruit juice concentrates have been very successful as colorants, even though they are primarily food ingredients. However, if a fruit juice contains sufficient pigment to be effective as a colorant, it is subject to the specifications of 21 CFR 73.250, which states that the colorant must result from the juice of mature edible fruits or be a water infusion of dried fruit. The colorant formulation can be either concentrated or dried. One product from grapes has been very successful. Since grape juice is one of the major components of fruit drinks, it is logical to sell both a compatible flavor and a colorant in the same formulation. Accordingly, grape varieties with more than eight times the normal concentration of anthocyanins are used to provide a flavorful red colorant. Yellow grapes are used to provide a yellow colorant. Since less processing is required, as compared with that necessary to purify the colorant, less pigment degradation occurs, thus providing a more attractive color.

A number of other fruit juice concentrates contain sufficient pigment to be effective colorants. These include elderberry, raspberry, strawberry, cranberry, orange, and tangerine. Others, such as acerola and guava, may contain a colorant but are usually added in such small proportions that they have little effect on color. Still others, such as apple, pear, grapefruit, banana, pineapple, papaya, mango, passion fruit, and kiwi, are simply classified as food ingredients.

Vegetable juice extracts are subject to the same specifications as described above for fruit juice extracts. Only two are important commercially—those prepared from red beets (21 CFR 73.40 and 73.260) and from red cabbage, which is permitted as an ingredient. Chapter 7 contains more information about these two colorants.

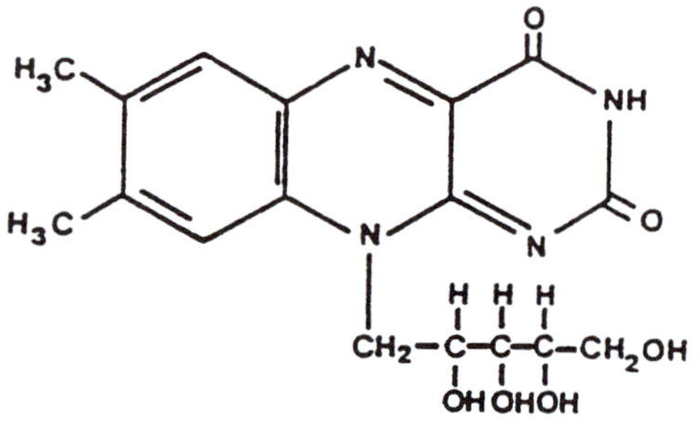

RIBOFLAVIN

Riboflavin (EEC No. E 101), a yellow pigment in the vitamin B group, is found in plant and animal cells. Originally extracted from natural products, it is now produced by synthesis. Riboflavin is usually added to food as an essential nutrient, but it is also an effective colorant. Riboflavin and riboflavin-5'-phosphate (Fig. 11-1) are well-known colorants in Europe, but only riboflavin is allowed in the United States. In pure form, it is a yellow-orange crystalline powder, soluble in water and alcohol and insoluble in ether and chloroform. Riboflavin is somewhat unstable to light and oxidation and is unstable in alkaline media. It is stable to heat. Riboflavin-5'-phosphate is more soluble in water, less bitter tasting, and more unstable to light. For colorant purposes, riboflavin is confined mainly to cereals, dairy products, and sugar-coated tablets. It produces an attractive greenish yellow color with an intense green fluorescence.

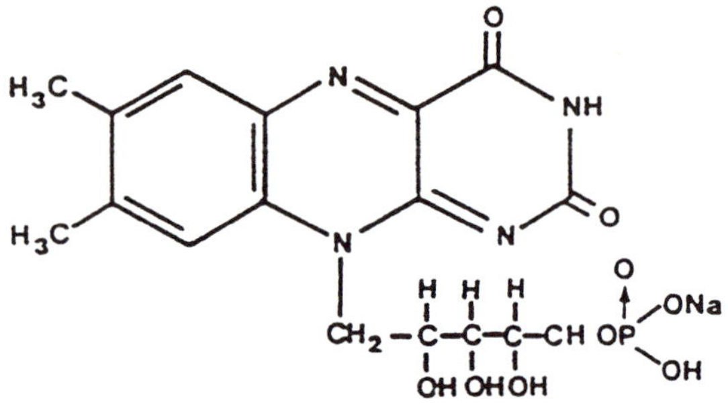

CORN ENDOSPERM OIL

Corn endosperm oil (CAS No. 977010 506) is obtained by extraction of corn gluten with isopropyl alcohol and hexane. It is a reddish brown liquid containing fats, fatty acids, sitosterols, and carotenoid pigments and their degradation products. In the United States, it is permitted in chicken feed to enhance the color of chicken skin and eggs.

ALGAE PRODUCTS

Dried algae meal. This is a dry mixture of alga cells produced by the growth of *Spongiococcum* spp. on molasses or corn-steep liquor. It contains a maximum of 0.3% of the antioxidant ethoxyquin. In the United States, it is permitted in chicken feed to enhance the yellow color of chicken skin and eggs.

Brown extract of algae. Algae, Brown Extract (CAS No. 977026 928) is produced by extraction of the seaweeds *Macrocystis* and *Laminaria* spp. It is not listed in the United States.

Red extract of algae. Algae, Red Extract (CAS No. 977090 042) is produced by extraction of the seaweeds *Porphyra* spp., *Gloiopeltis furcata*, and *Rhodomenia palmata*. It is not listed in the United States.

COTTONSEED PRODUCTS

The FDA recognizes four products from cottonseed kernels: cottonseed flour CAS No. 977050 546, which is partially defatted and cooked; cottonseed flour CAS No. 977043 778, which is partially defatted, cooked, and roasted; cottonseed kernels CAS No. 997043 778,

Fig. 11-1. Structures of riboflavin and riboflavin-5'-phosphate.

which are glandless raw kernels; and cottonseed kernels CAS No. 997043 78, which are glandless roasted kernels. Under 21 CFR 73.140, only "toasted, partially defatted, cooked cottonseed flour" is permitted in foods in the United States. These products are usually considered to be food ingredients and contribute only marginally to a yellow color.

SHELLAC

Shellac is obtained from the resinous secretions of the insect *Lassifer lacca*. Bleached shellac is produced by dissolving the raw lac in aqueous sodium carbonate solution, followed by bleaching with sodium hypochlorite. The bleached lac is precipitated with dilute sulfuric acid and dried. It is an off-white, amorphous granular resin, slowly soluble in ethanol, insoluble in water, and slightly soluble in acetone and ether. Bleached shellac is usually dissolved in an appropriate solvent before application to foods or food wrappers. Bleached wax-free shellac is manufactured in a similar manner except that a filtration step is introduced after the initial solubilization to remove the wax.

Shellac is used primarily in foods as a coating agent, surface-finishing agent, and glaze in baked goods and some fruits, vegetables, and nuts. It is not primarily a colorant but does affect the appearance of a product. The FDA (4) listed 10 entries for shellac.

OCTOPUS INK AND SQUID INK

The secretions of the octopus and the squid are complex mixtures of melanoidin polymers. They constitute an effective black colorant for pasta for special occasions in some ethnic groups, particularly the Portuguese. However, it is not a permitted color additive in the United States.

References

1. Francis, F. J. 1996. Safety of natural food colorants. In: *Natural Food Colorants*, 2nd ed. G. A. F. Hendry and J. D. Houghton, Eds. Blackie Publishers, Glasgow, Scotland.
2. Marmion, D. M. 1991. *Handbook of U.S. Colorants*, 3rd ed. John Wiley & Sons, New York.
3. National Academy of Sciences. 1981. *Food Chemicals Codex*. National Academy Press, Washington, DC.
4. U.S. Food and Drug Administration. 1993. *Everything Added to Food in the United States*. CRC Press, Boca Raton, FL.

CHAPTER 12

Baked Goods, Cereals, and Pet Foods

In This Chapter:

Baked Goods and Cereals
  FD&C Colorants
  Natural Colorants
  Shades

Pet Foods
  Water-Soluble Colorants
  Lakes
  Exempt Colorants

Troubleshooting

The first part of this chapter, on baked goods and cereals, contains information about FD&C colorants and natural colorants and a section on shades, in which information from both sections is covered. The second part discusses these colorants in the context of pet foods.

Baked Goods and Cereals

Several factors must be considered in the choice of a suitable colorant for a particular baked or cereal product. The fat content of some bakery goods can present special problems, so oil-dispersible FD&C lakes or oil-soluble natural colorants may be required. Colorants chosen for cereals must be heat and extrusion stable—a limitation for some of the natural colorants. In addition, colorants (natural and FD&C) cannot be added to products that currently have a *standard of identity* unless the standard calls for the use of specific colorants.

FD&C COLORANTS

The appeal of cakes and sweet rolls can be improved by the judicious use of color. For this application, the liquid market form is recommended. The liquid colorant is generally diluted before addition to dough, and it is best to keep the colorant solutions fairly dilute. For most bakery products, 30 g of full-strength colorant per liter of water (4 oz/gal) is ample, and, for very light shades, it is often better to work with a colorant strength of 15 g/L (2 oz/gal). With such solutions, care must be taken to allow sufficient time for uniform color mixing in each batch.

Colorants and fat. Cookie fillings and bakery coatings represent special problems because of the presence of shortening. Cookie fillings can usually be colored by gradually adding liquid colorant prepared with propylene glycol. Mixing the colorant and filling thoroughly is very important. Also, during preparation of a liquid solution, the solution must be filtered to remove sediment that may cause specks in the finished product.

The fat content of bakery coatings varies considerably. Therefore, no single colorant procedure yields satisfactory results in all situations. Since most synthetic food colorants are water soluble, some method must be found to reconcile the soluble colorants with the fat

Standard of identity—A legal standard, maintained by the FDA, that defines a food's minimum quality, required and permitted ingredients, and processing requirements, if any. It applies to a limited number of staple foods.

in the bakery coating. This can usually be done by using one of the following: glycerin solution, propylene glycol solution, mixed glycerin and propylene glycol solution, any of the above plus lecithin, or FD&C lake pigments. Try the simplest approaches first, e.g., coloring with FD&C lake pigments (dispersed in oil or other suitable carrier) or with water-soluble colorants dissolved in glycerin or a propylene glycol solution.

In general, glycerin is preferred over propylene glycol as a solvent. Because most soluble colorants have a higher solubility in glycerin, solutions with higher concentrations can be prepared if glycerin is used. The only disadvantage of glycerin is the somewhat higher price. However, this is a minor consideration because colorant represents a small fraction of the total cost of goods sold, while the color effect constitutes a significant percentage of the product's appeal.

When using lake pigments in baker's sugar icings and fondant icings, the FD&C lake can be added to the sugar or water portion before blending the mixture in a low-speed, high-shear mixer. About 0.63 g of colorant per kilogram of icing (1 oz/100 lb) produces soft pastel shades that are stable to light. Lakes are recommended because of their stability in light and their nonmigration properties.

For oil-based coatings, lakes can be incorporated into the vegetable fat portion before the remaining ingredients are added. A good mixer successfully disperses the lakes into the fluid coating. Although it is not necessary, predispersion of the lakes in a hydrogenated oil is recommended. Using 0.63 g of lake per kilogram of coating (1 oz/100 lb) results in attractive pastel shades that simulate fruit colors and shades such as caramel, butterscotch, and chocolate.

For sandwich cookie fillings, lakes are more suitable than water-soluble colorants because lakes disperse better in the fat portion of the fillings. Water-soluble colorants are incompatible. The lakes can be mixed with the sugar before the shortening is added. Mixing 3.13 g of colorant per kilogram of filling (0.5 oz/100 lb) produces desirable shades.

When using water-soluble colorants in the applications described above, add the liquid colorant slowly and stir the melted coating vigorously. Then observe the results.

Solution concentration. Two things should be considered in deciding upon the concentration of a colorant solution. First, if the addition of liquid to the material must be held to a minimum, a highly concentrated solution is indicated, which, for most colorants, requires glycerin as the solvent. Second, the shade of the final product may vary as a result of concentration; that is, equivalent amounts of colorant introduced with different solvent volumes do not necessarily produce equivalent shades in the final product.

The second consideration is particularly important when a more concentrated colorant solution is being prepared. In this situation, it is imperative that laboratory samples of the finished product be prepared so that the correct volume of the new colorant solution can be accurately determined.

Lecithin. In extreme cases in which the fat content of a coating is so high that even glycerin solutions do not yield satisfactory results, specking and uneven coloring can often be corrected by adding a small amount of lecithin. Mixtures of lecithin and propylene glycol solutions have also produced good results. The best solution, however, is the use of lakes in an edible carrier such as vegetable oil.

Dry mixes. Dry cake and doughnut mixes of flour and other ingredients can be subtly colored to enhance their appearance. Orange blends, brown shades that simulate cinnamon and spices, and yellow shades for butter and egg can be blended with all or part of the flour portion before the other ingredients are added. For better dispersion, the premix should be milled. A common usage is 0.63 g of lake or lake blend to 1 kg of mix (1 oz/100 lb).

Sugars. Flavored frosting sugars for coating doughnuts and other pastries are colored with water-soluble colorants, lakes, or sometimes a combination of both. Colored sugar is used principally for decoration, so it is not necessary to carry the colorant forward into another product. Liquid is the market form of choice. For some colored sugar applications, however, lake blends are preferred.

When sugar is colored, a colorant solution is sprayed into the sugar as it is being mixed. Mixing is then continued for a short time until the colorant is uniformly distributed. If the solution is concentrated (about 60 g of colorant per liter [8 oz/gal]), only a small volume is required and not much water is introduced. The sugar can be air dried at room temperature or dried in a gentle current of warm air. Although the sugar has a slight tendency to cake during drying, it can be easily broken up to become free flowing again.

Water-soluble colorant and lake combinations can be used to produce bright decorating-sugar crystals. The colorant is added to sanding (coarse-granulated) sugar in a coating pan with a small amount of water or simple syrup to develop the crystal color. An easier method is to use lakes and water-soluble colorants dispersed in syrup.

NATURAL COLORANTS

Natural colorant systems (individual colorants or combinations) are commonly used in baked goods and cereals in the United States.

Sweet goods are commonly colored with water or oil-soluble annatto extract or annatto and turmeric blends to yield pleasing butter-to-egg-yolk shades. Usage rates vary depending on the strength of the colorants. A typical usage rate would be 0.02–0.06%, depending on the product and the desired final shade.

Depending on the flavor and the desired shade, crackers are commonly colored with annatto extract, turmeric oleoresins, paprika oleoresins, caramel colorants, or various combinations of the above. A cheese-flavored cracker that a product developer would like to color an "orange, cheddar cheese" shade would require an emulsion-type or water-soluble annatto extract.

Cereals can be colored with a variety of natural colorants, depending on the application and the desired final shade. Typically, annatto extract and turmeric oleoresin are used. When FD&C Yellow No. 5 was the only certified colorant required to be labeled on ingredient statements, which could be seen as a negative by "natural"-minded consumers, many food companies chose to use certain forms of turmeric oleoresin instead (turmeric exhibits a bright lemon yellow shade almost identical to that of FD&C Yellow No. 5). Even though all FD&C colorants now must be specifically declared on ingredient statements, turmeric is still extensively used in the cereal industry.

In cereal, pigment forms of turmeric (instead of polysorbate 80-based turmerics) are used. In the pigment form, turmeric colors through dispersion until heat is applied. Extrusion temperatures develop turmeric from a pale yellow-orange to an intense, bright lemon yellow. Turmeric is often combined with FD&C Blue No. 1 or FD&C Yellow No. 6 to yield green and orange shades, respectively. Water-soluble annatto extract has been successfully used to yield golden cereal shades, and, where a magenta red shade is desired, carmine-type colorants are suitable for cereal applications.

SHADES

In general, various appealing shades for baked goods and cereals can be achieved by using the following color systems.

Red shades. *FD&C Red No. 40.* This orange-red shade is generally extrusion stable. The colorant is most often incorporated as a liquid. The lake form is appropriate for fat-based phases or to add colorant to a dry mix.

FD&C Red No. 3. This bright, clean red is referred to as bluish red in hue. The lake form was delisted in 1989. The colorant is appropriate for pink shades and yields excellent purple shades in combination with FD&C Blue No. 1 or 2.

Carmine and related products. These products are available in lake and water-soluble forms and are the most stable of all the natural reds.

Cochineal extracts. These water-soluble extracts are useful for shade variations. The shade is dependent on pH; acid-stable forms that are red at neutral pH are available.

Beet juice products. Beet juice products have poor heat stability (will lose color and change shade) but may work in some cereal applications, such as in a cocoa-based product in which browning is desirable or in a sugar-type coating for cereal pieces.

Anthocyanins. Because anthocyanins are not heat stable (red cabbage is the most stable) and are red only at a pH below 3.8, they are not useful in most cereal applications. However, they may work in a sugar-type coating for cereal pieces.

Orange shades. *FD&C Yellow No. 6.* This bright orange shade is often blended with FD&C Yellow No. 5 to yield butter-to-egg-yolk shades. It is extrusion stable.

Paprika oleoresin. Paprika oleoresin is naturally oil soluble, but water-dispersible forms may be applicable. However, without an antioxidant to protect the pigment, it will fade during heat processing. Flavor may also be an issue, depending on the amount used.

β-Carotene. β-Carotene is available in water-dispersible forms, but large amounts are required (small amounts yield yellow shades). It is heat stable.

Canthaxanthin. This water-dispersible, synthetic carotenoid is heat stable. However, its low usage restriction of 30 mg/lb [66 mg/kg] of food may make it unsuitable for use in cereal.

β-Apo-carotenal. This colorant is oil soluble; no water-dispersible form is commercially available in the United States. Its low usage restriction of 15 mg per pound [33 mg/kg] of food may make it unsuitable for use in cereal.

Orange shades via blends. The following combinations produce acceptable orange shades: FD&C Red No. 40 and FD&C Yellow No. 5; carmine and turmeric products (various ratios for various shades); carmine and annatto; and carmine and β-carotene.

Yellow shades. *FD&C Yellow No. 5.* This bright lemon yellow shade is extrusion and oven stable.

Turmeric products. These bright lemon yellow shades are available as powders, dispersions, and blends. They have excellent heat stability but poor light stability. The pigment form used most often (in powder or dispersion form) gives the dough a light yellow color that intensifies to lemon yellow during the extrusion process.

β-Carotene products. These products have excellent heat stability, but the color must be protected with an antioxidant. They are available in water-dispersible forms.

Blue shades. *FD&C Blue No. 1.* This stable, bright blue (slightly yellow) shade is commonly used in caramel, nut, and chocolate shades in combination with other approved colorants. The usage and consumption levels must be considered to avoid discoloration of the stool.

FD&C Blue No. 2. This flag-blue shade is not stable as a liquid colorant but is extrusion and oven stable. It is often used in combination with other approved colorants to yield caramel, nut, and chocolate brown shades in a variety of baked goods and cereal applications.

Natural blue. No natural blue colorant is currently permitted by FDA as a color additive for foods in the United States. Natural blues do exist in nature. One of the first products people think of as a possible source of natural blue is blueberries. Unfortunately, blueberry color is red upon extraction. Phycocyanin is available from blue-green algae, but it is permitted only in Japan and the cost is approximately 200 times that of FD&C blues. Anthocyanins from sources such as red cabbage, elderberry, and grape juice provide blue to purple shades but only at pH levels above 3.8, where their stability is compromised. Grape Skin Extract is permitted in the United States only as a colorant for beverages. Blue shades for cereal are typically achieved by using FD&C Blue No. 1, FD&C Blue No. 2, or combinations of the two.

Green shades. No natural green colorant is currently permitted by FDA for use in foods, and no natural blue is available to blend with yellow to obtain a green. Natural green juices are available, from spinach and other green vegetables, for example, but these colorants have poor stability, and there are issues concerning concentration and flavor. Sodium copper chlorophyll is permitted only in dentifrices. Green colorants for cereal applications are typically achieved by combining FD&C Blue No. 1 with turmeric or FD&C Yellow No. 5.

Pet Foods

The U.S. food regulations state that colorants added to pet foods must be the same as those approved for use in food consumed by humans, with the exception of iron oxides. Soluble colorants are extensively employed because they are easy to use; a wide range of colors is available; and there are no usage limitations. In addition, titanium dioxide and lakes are often used, depending on the particular pet food application. On the other hand, iron oxides, while widely used in pet food, are more difficult to use, are limited to neutrally colored products, and have a usage limitation of 0.25% in the United States. Internationally regulated food colorants are also used in pet food applications outside the United States.

WATER-SOLUBLE COLORANTS

Solutions of soluble colorants are the preferred form because of the additional control they provide. It is possible, however, to use powder if the pet food contains sufficient moisture to dissolve the dry colorant. If dry colorants are used, it is important to ensure that they completely dissolve and disperse. One application in the pet food industry that requires special attention is gravy color coatings. Nonflashing blends of soluble colorants are blended with flavors and thickening agents. This mix is then "dusted" onto the surface of the feed. When the consumer adds water, the ingredients combine to create an appealing instant gravy. If nonflashing blends are not used, the consumer will see the initial "flashing" of red, yellow, and blue primary colorants as they dissolve. Furthermore, in areas of high humidity, these surface applications can develop an unappetizing green hue.

LAKES

Pet food manufacturers have benefited greatly from the introduction of FD&C lakes. Because lakes are not affected by high temperatures or protein interaction (both of which cause fading of the FD&C water-soluble colorants), they have been especially useful in canned pet food products, for which retorting is necessary. Lakes also perform especially well in semimoist pet foods such as burger types, where extra stability is required because the product is displayed through a window in the package. The concentration of lakes used depends upon the background color of the grain and meat used and upon the shade desired. As a starting point, 0.05–0.2% of lakes or lake blends can be tried.

EXEMPT COLORANTS

Titanium dioxide, the only white pigment permitted by FDA, is the most commonly used colorant in pet food applications, mainly because of its hiding power or tinting strength. Often the desired red (meat) shade cannot be achieved with FD&C Red No. 40 alone because of the base color of the grains or meat by-products used. Titanium dioxide can be used at a rate of up to 1% of the finished weight of the product as consumed. At levels up to this generous usage restriction, opacification and coverage of the base shade can be achieved. FD&C Red No. 40 can then be added, and the desired appealing bright red shade can be attained. A similar strategy, with titanium dioxide and FD&C Yellow Nos. 5 and 6, can be employed for yellow-to-orange "cheese" or "egg" pieces.

Other colorants exempt from certification are used in pet foods, but because of the increased cost associated with their use, they are not common. Most desirable shades can be achieved with the certified colorants. Several "natural" dog biscuits colored with beet juice, turmeric oleoresin, and annatto are presently on the market. These up-scale biscuits appeal to a certain market niche.

Natural colorants that would be stable alternatives for retort products include carmine, β-carotene, and turmeric.

Troubleshooting

BAKED GOODS AND CEREALS		
Symptoms	Causes	Changes to Make
Color not uniform throughout product	Insufficient mixing	Increase mixing time. Change point of addition to allow more mixing time.
	Colorant doesn't mix with product primary phase	Use a colorant carrier that will mix with product primary phase.
	Colorant present in very small quantity	Use a more dilute colorant preparation.
Specks in finished product	Precipitation or sediment in colorant preparation	Filter colorant before using.
	Colorant incompatible with product pH	Change to colorant compatible with pH or use pH-stabilized version.

Symptoms	Causes	Changes to Make
Color of final product too light, changes in hue, or disappears	Colorant concentration too low	Use more colorant.
	Colorant sensitive to processing temperatures	Use a more heat-stable colorant. Increase colorant dosage to compensate. Adjust processing conditions to minimize exposure to high temperatures. Adjust point of addition of colorant to minimize exposure to high temperatures.
	Colorant sensitive to oxygen incorporated during processing	Use a more oxygen-stable colorant. Minimize oxygen incorporation during processing. Use an antioxidant in conjunction with colorant.
	Colorant-ingredient interaction	Look for negative interactions with salts, proteins, or fortification ingredients.
	Selective fading of one component of a colorant blend	Reformulate blend to remove less-stable component. See suggestions above for stabilizing colorants.
Color of final product too dark	Colorant concentration too high	Use less colorant.
Color of final product dull	Background color from other ingredients interferes with colorant	Take steps to minimize background color from other ingredients.
Final product color fades with time on shelf	Light-induced degradation of colorant	Use a more light-stable colorant. Use an antioxidant in conjunction with colorant. Eliminate clear window on package. Use a UV-absorbing film on window.

PET FOODS

Symptoms	Causes	Changes to Make
Color is not uniform, specking	Dry colorant not dissolving properly	Predissolve colorant in some of the process water. Use a liquid colorant preparation.
	Insufficient mixing	Increase mixing time. Change point of addition to allow more mixing time.
	Colorant doesn't mix with product primary phase.	Use a colorant carrier that will mix with primary phase.
	Colorant present in very small quantity	Use a more dilute colorant preparation.
Primary colors are seen as the liquid is added to make gravy	Colorant flashes	Use nonflashing colorants.
Color fades during processing	Colorant is sensitive to processing temperatures	Use a more heat-stable colorant. Use FD&C lakes instead of dyes. Increase colorant dosage to compensate. Adjust process to minimize exposure to high temperatures. Adjust point of addition of colorant to minimize exposure to high temperatures.

Color of final product too light, changes in hue, or disappears	Colorant concentration too low	Use more colorant.
	Colorant sensitive to oxygen incorporated during processing	Use a more oxygen-stable colorant. Minimize oxygen incorporation during processing. Use an antioxidant in conjunction with colorant.
	Colorant-ingredient interaction	Look for negative interactions with salts, proteins, or other ingredients.
	Selective fading of one component of a colorant blend	Reformulate blend to remove less-stable component. See suggestions above for stabilizing colorants.
Color of final product too dark	Colorant concentration too high	Use less colorant.
Color of final product dull	Background color from other ingredients interferes with colorant	Take steps to minimize background color from other ingredients.
Final product color fades with time on shelf	Light-induced degradation of colorant	Use a more light-stable colorant. Use FD&C lakes instead of dyes. Use an antioxidant in conjunction with colorant. Eliminate clear window on package. Use a UV-absorbing film on window.
Dairy-based brown-colored product turns green	Microbial attack on azo dye (FD&C Red No. 40), leaving remaining components yellow and blue	Use FD&C Red No. 3, carmine, or other red colorant resistant to microbial attack.
Color of "meat" not correct	Interference from color of grain	Use titanium dioxide to mask grain color.

CHAPTER 13

Beverages and Dairy Products

Beverages

Beverages represent one of the largest markets for both natural and FD&C colorants. The popularity of dry-mix beverages, still beverages (including new-age types), and carbonated soft drinks continues to grow.

Historically, carbonated and most still and dry-mix beverages have used FD&C colorants to yield appealing, bright, and typically stable shades. Because of the cost and issues related to stability, natural colorants were not chosen for beverages, with the exception of caramel colorants, which have been used to color the cola and root beer types of carbonated beverages. However, use of natural colorant in beverages is now increasingly popular, especially in the "new-age" (juice-containing) beverages. Increased availability, globalization, and improved extraction and concentration techniques have all contributed to the popularity of natural colorants.

FD&C COLORANTS

Water-soluble colorants are the obvious choices for almost all beverage applications. Lake pigments, which are appropriate for dry-mix beverages, are discussed later in this section.

To color carbonated beverages, the colorant is often combined with the flavor in a product known to the trade as a "compound." When an FD&C colorant is dissolved with flavor in a compound, solubility problems occur. Even though there may be enough water in the compound to keep the colorant in solution under ordinary circumstances, the presence of water-miscible solvents associated with flavor oils, such as alcohol and propylene glycol, can reduce the solubility of some colorants. If a large proportion of the liquid in the compound consists of such solvents required to keep flavor oils in solution, it may be impossible to dissolve the required quantity of colorant.

If solubility problems arise, five alternatives are generally available: 1) substitute glycerin for propylene glycol or alcohol as a flavor oil solvent, 2) change the colorant blend to one that contains primary colorants with high solubility, 3) reduce the strength of the flavor, 4) pack the colorant as a separate unit to be dissolved by the bottlers and added to their syrup, and 5) try a darker shade that permits use of less colorant.

In This Chapter:

Beverages
  FD&C Colorants
  Natural Colorants

Dairy Products and Spreads
  Butter
  Margarine
  Spreads
  Yogurt
  Fat-Based Coatings
  Retorted Milk Products

Troubleshooting

Another potential problem in carbonated soft drinks is color fading caused by ascorbic acid, which is added as a vitamin source and an oxygen scavenger. Products containing high levels of ascorbic acid must be bottled immediately. The shelf life will be short, and factors such as exposure to sunlight must be minimized. If the ascorbic acid is added only as an oxygen scavenger, the head space should be reduced so that not more than 10 mg of ascorbic acid is required per bottle. The colors of bottled beverages usually fade as the flavor degrades.

No fading problem occurs with carbonated beverages packaged in cans. However, can corrosion is a possibility. Soluble colorants themselves are not corrosive, but they can interact with other ingredients and carbon dioxide to attack the metal. Corrosion can be minimized to acceptable limits by reducing the concentration of azo colorants to 50 ppm or less. FD&C Blue No. 1 (Brilliant Blue) and FD&C Green No. 3 (Fast Green FCF) apparently do not contribute to corrosion.

Can corrosion is not a problem with still beverages because no carbon dioxide is present, but the problem of color fading can occur. Fading can be significantly reduced by fruit juice in the beverage.

In drink powders for which pronounced color in the dry mix is required, the usual procedure is to blend plating-grade powder with other ingredients. Thorough mixing is essential if the dry product is to appear uniform. If a more intense color is desired, a small amount of colorant can be atomized onto the dry mix. Under certain circumstances, water-soluble and lake combinations can be used. Lake pigments give the dry powder the desired color and, depending on the usage level and pH of the final product, may contribute color to the final liquid beverage. If the pH is low enough (below 3.5), the lake breaks down into the free pigment. If the pH of the final beverage is above 3.5, the lake contributes some cloudiness or opacity because of its insolubility.

NATURAL COLORANTS

Natural colorants can be successfully used to provide appealing, natural-looking shades in a variety of beverage applications. However, certain issues regarding the use of natural colorants must be considered, e.g., regulations; technical properties of the various colorants, including stability under different conditions of heat, light, pH, and levels of ascorbic acid; and interactions with other beverage ingredients.

Although many natural colorants exist, some, such as titanium dioxide, are not applicable to use in beverages. Others, such as turmeric oleoresin and paprika oleoresin, may be used in frozen concentrates and sometimes in canned concentrates or mixes. Annatto extract, β-carotene, cochineal extract and carmine, vegetable juice, fruit juice, grape skin extract, caramel, and canthaxanthin are commonly used colorants in beverages; their technical properties are discussed in the following sections. Typical usage levels (which depend on the concentration of the color itself and the desired shade) are given in Table 13-1.

Annatto extract. Annatto extract is a red-orange colorant. When the colorant is extracted with an alkaline system such as potassium hydroxide, the oil-soluble bixin pigment is converted into the water-soluble norbixin pigment. Typical alkaline annatto extracts such as this are not soluble in systems with a pH below 4.0; therefore, the norbixin pigment precipitates in these systems. However, "acid-proof" forms, containing suitable emulsifiers or starch, allow the norbixin to remain soluble or dispersible in systems with low pH.

Annatto extract has fair to good heat stability. It will survive hot-fill system requirements, typically losing only 8–10% of its initial color. Its light stability is poor to fair, and it is currently being used in clear-pack beverages with varying degrees of success. If it is used at low levels (0.005–0.007%), the product fades more quickly than if it is used at higher levels (0.03–0.06%). Depending on the usage rate, a yellow-orange to red-orange shade can be achieved.

β-Carotene. As a colorant, β-carotene exhibits a yellow to orange shade, depending on the product and quantity used. Typical orange juice shades can be achieved in beverage applications.

β-Carotene is naturally oil soluble. Water-dispersible preparations are also available and contribute cloud when used in water-based applications such as beverages. Also, because of its oil-soluble nature, β-carotene tends to migrate to any free oil phase, yielding a colored "ring" in some beverage applications. To minimize the chance of this ring forming, one must be sure that any flavor oils or clouding agents used are homogeneously blended with the other ingredients.

β-Carotene exhibits excellent heat and light stability and is stable in acid systems. In a hot-fill beverage, it fades only slightly (7–10%) during the initial heat processing and then only slightly, if at all, during accelerated storage conditions (40°C, 104°F). It is susceptible to oxidation, and colorant preparations must be protected by the addition of an antioxidant. β-Carotene is stabilized by the presence of ascorbic acid.

Cochineal extract and carmine. Cochineal extract (carminic acid) exhibits shade changes with changes in pH. At pH levels of 4.0 and below, it is orange; at 4.0–6.0, it is a magenta red color; and above 6.0, it is a blue-red shade. An "acid-proof" form is available, in which the color remains red at pH 2.7 and above. Cochineal extract is commonly used to color beverages—from bright, stable orange shades to bright magenta red shades. The aperitif Compari is colored with cochineal extract.

Carmine, a very stable colorant, is commonly referred to as the lake (pigment) form of cochineal extract. Carminic acid readily binds with various ions, including aluminum and calcium. Various shades

TABLE 13-1. Typical Usage Levels for Natural Colorants

Colorant	Usage Level (%)
Annatto extract	0.04–0.08
β-Carotene	0.05–0.10
Cochineal extracts	0.05–0.15
Carmine liquid colors	0.05–0.10
Vegetable juice concentrates	0.03–0.06
Fruit juice concentrates	0.05–0.15
Grape skin extract	0.05–0.15
Caramel color	0.10–0.25

Common Exempt Colorants Used in Beverages with pH <3.5

Acid-proof annatto extract

Canthaxanthin

Caramel

Carmine (if pulp present)

β-Carotene

Cochineal extract (orange and red)

Elderberry juice concentrate

Grape skin extract

Red cabbage juice

> **Common Exempt Colorants Used in Beverages with pH >4.0**
>
> Annatto extract
> Canthaxanthin
> Caramel
> Carmine liquid colorants
> β-Carotene
> Cochineal extract

from yellow-red to magenta-red to violet-red can be obtained. Carmine lake is often solubilized with alkali (sodium, ammonium, or potassium hydroxide) to yield a variety of water-soluble liquids commonly used to color beverages. Alkaline solutions of carmine precipitate at pH 3.5 and lower. This is used to advantage in certain applications such as pulp-containing beverages. The precipitation can be avoided by keeping the pH of the system above 3.5.

Cochineal extract and carmine have good to excellent heat and light stability, depending on the ingredients. In hot-fill beverage systems (such as a beverage containing 5% juice), an acid-proof form of cochineal extract does not lose any strength when heated to 82.2–85°C (180–185°F) for 5 min. Over time, cochineal extract loses some of its "blue" notes, becoming more orange-red. In a recent storage study (1), this change in shade was stabilized by the presence of ascorbic acid (250 ppm).

Cochineal extract and carmine yield excellent, very appealing grape shades in combination with FD&C Blue No. 1 and FD&C Blue No. 2 lake as appropriate. Often cochineal extract and carmine (and other "natural" colorants) are used because of the characteristic shade and not because they are "natural."

Vegetable juice concentrates. The two vegetable juices most commonly used as colorants are from red cabbage and beet. The beet juice is not stable in beverage systems.

Red cabbage juice concentrate. Red cabbage juice concentrate is anthocyanin based. Like all anthocyanins, those in red cabbage juice change shade with a change in pH and are most stable in systems with low pH. The anthocyanins of red cabbage exhibit good stability in terms of heat and light. Red cabbage juice is commonly used in beverage applications at concentrations of either 0.007 or 0.05%. At these levels, the color survives the heat-processing requirements of hot-fill beverages; 80 and 100% of the original color remains, respectively. A recent study of long-term storage of beverages colored with red cabbage juice at both rates (1) showed that only 20% of the color was lost at a storage temperature of 40°C (104°F) for three weeks. Shelf life decreased more dramatically in the presence of ascorbic acid, especially when the lower amount (0.007%) of colorant was used to achieve a "pink lemonade" shade. At this level and with ascorbic acid, only 50% of the color remained in a lemonade beverage (Table 13-2) after three weeks at 40°C.

Beet juice concentrate. Beet juice is most stable in foods with low water activity, such as frozen novelties and fruit fillings. It is not stable in liquid systems such as beverages—it does not survive the hot-fill process and becomes brown and less intense. In products that do

TABLE 13-2. Formula for Lemonade (5% Juice) Used in Stability Studies

Ingredient	Percentage
Purified water	83.271
Ascorbic acid	0.025
Universal Flavors No. 404551 lemonade flavor syrup	16.654
Colorant (typical usage)	0.050

not undergo heat processing, beet discolors and becomes brown within a week on the shelf.

Dry-mix beverages (clear and opaque) are appropriate for beet powder, which has good to excellent stability in these applications. The final liquid beverage is typically consumed before severe degradation of the beet juice occurs (three days, refrigerated).

Fruit juice concentrates. Fruit juices as color additives must meet the specifications of 21 CFR 73.250, which states that the colorant must result from expressing the juice from mature varieties of edible fruits or be a water infusion of dried fruits. Fruit juices that are concentrated enough to be used as color additives include those from elderberry and grape.

Elderberry juice concentrate. This anthocyanin-based colorant is commonly used in beverages to produce an appealing "berry" shade. Elderberry anthocyanins are much redder and less blue than those from sources such as red cabbage and grapes. Unfortunately, the long-term heat stability of elderberry juice is less than that of red cabbage. A recent study (1) showed that although elderberry juice anthocyanins at a usage level of at least 0.08% survived the heat processing in a lemonade-based, hot-fill beverage system, long-term storage (three weeks at 40°C) resulted in a color loss of 38%. The loss was 72% in the presence of ascorbic acid (250 ppm).

Grape juice concentrate. Grape juice used at a typical rate of 0.15% yields appealing blue-red shades in acidic systems such as beverages. Grape juice anthocyanins, like all anthocyanins, are most stable at a pH below 3.8. This product survives in a lemonade-based beverage without losing strength during the heat-processing step. However, long-term storage under accelerated conditions (40°C for three weeks) resulted in a color loss of 39% (1). In the presence of ascorbic acid (200 ppm), 65% of the color was lost after three weeks of storage under the same oven conditions. Color intensity, however, was not the only change. The beverage color became less blue and more orange-red with a slight brown note.

Grape skin extract. Grape skin extract (GSE), also called enocianina, is a color additive permitted in 21 CFR 73.170 only for coloring beverages.

GSE contains anthocyanin pigments similar to those found in the red grape juice concentrate described above. When used at the typical rate of 0.08%, GSE anthocyanins survive the hot-fill beverage process in a lemonade-based beverage without losing any color intensity. The shade changes slightly, losing some blue notes. In a recent study (1), after long-term exposure to heat (40°C for three weeks), GSE retained 68% of its original color. The shade of the beverage became less blue and more yellow-red with a slight brown tinge. As expected, in the presence of ascorbic acid and exposed to the same storage conditions, the beverage retained only 50% of its original color strength. It became weak and developed a yellow-red shade with brown notes. The light stability of GSE is fair.

Caramel. In general, caramel colorants exhibit good heat and light stability. Acid-stable forms are available for use in beverages. In the United States, we are fortunate to have caramel as part of our natural color palette. It is currently the only way to obtain a "natural" brown shade, since we do not have any natural blues or greens available for blending with a yellow or red to obtain brown. Caramel is commonly used to color a variety of beverages.

Canthaxanthin. Canthaxanthin is a synthetically prepared carotenoid that is available in a water-dispersible form. It produces an orange-red shade when diluted, depending on the product and quantity used. Canthaxanthin is restricted in foods to 30 mg per pound of finished product. It is a stable colorant applicable to certain beverages.

Dairy Products and Spreads

Attractive tints can easily be produced in most dairy applications. Dairy products are covered in this chapter in terms of the most commonly used colors.

BUTTER

Butter has a standard of identity in the United States that allows for the addition of β-carotene, the naturally occurring colorant found in cow's milk. When butter is made, β-carotene concentrates in the oil phase and gives butter its characteristic yellow color.

MARGARINE

β-Carotene is the most commonly used color additive for margarine in the United States. Its oil-soluble nature and natural presence in butter make it the colorant of choice for margarine. In addition, because of its provitamin A activity, β-carotene contributes to the vitamin A content of margarine.

SPREADS

Most of the low-fat and no-fat margarine-type spreads on the market are colored with FD&C Yellow Nos. 5 and 6. Other water-soluble and dispersible colorants could also be used but not with as much success as this combination. Water-dispersible β-carotene and some forms of annatto extract can sometimes result in a slight pinking reaction, depending on the nature of the ingredients of the spread itself.

YOGURT

Yogurt can be colored with water-soluble colorants, either certified colorants or exempt (natural) colorants, depending on the ingredients and processing used. For many berry flavors in active-culture yogurts, cochineal extract, carmine, beet juice, and FD&C Red No. 3 can be utilized successfully, alone or in combination with other water-soluble colorants.

FD&C Red No. 40 is not commonly used in active-culture dairy products because of microbiological degradation. Microorganisms break the azo bond in FD&C Red No. 40, causing it to become colorless. This undesirable property is magnified when other, more stable colorants such as FD&C Blue No. 1 and a yellow component are present (in a chocolate flavor, for example). In this example, when FD&C Red No. 40 is decolorized, an undesirable green shade remains.

Water-soluble carmine, cochineal extract, beet juice, and FD&C Red No. 3 are resistant to microbial attack. However, because of possible future legal action against FD&C Red No. 3, carmine, cochineal extract, and sometimes beet juice are more commonly used.

In yogurt applications, the colorant is often added to a fruit preparation, which is then combined with the yogurt (either blended with the yogurt or placed as a separate phase in the cup to be blended before consumption). The low pH of the fruit preparation dictates that an acid-stable, water-soluble colorant be added. This precludes the use of FD&C lakes and carmine lake (which break down at low pH into the free pigment component) and FD&C Red No. 3 (which is insoluble in systems with a pH of less than 4.0).

Beet juice colorant is typically added to the yogurt itself and not to fruit fillings (especially aseptically packaged fruit fillings) because heat processing degrades beet juice pigments.

The most commonly used red colorants for yogurt are cochineal extract and water-soluble carmine products, which are stable during heat processing, resist microbial attack, and yield very bright, appealing shades. The form of cochineal extract needs to be considered. Cochineal extract direct from the cochineal insect results in an orange shade at a pH below 4.0, although it is a magenta red shade at a higher pH. Typically, cochineal extract that is considered "acid proof" is used. These cochineal extracts remain red from low to high pH (2.8–7.0).

FAT-BASED COATINGS

For ice cream novelties with hard fat coatings, lakes are important coloring agents. Pure white coatings can be used and tinted many attractive shades with FD&C lakes and carmine. These pigments should be dispersed in vegetable oil before they are incorporated into the coating to ensure maximum dispersion of the pigments in the coating. For best results, a predispersed preparation from a colorant supplier should be used.

RETORTED MILK PRODUCTS

Because FD&C Red No. 40 does not hold up under retorting in calorie-controlled dietary liquids, water-soluble carmine products, FD&C Red No. 3, or FD&C lakes are often incorporated. If lakes are used, they should be homogenized into the liquid. In most formulations, lakes do not settle out or separate but remain suspended. If settling may occur, the product should be labeled "Shake well before using." As a starting point in evaluating lakes for this application, begin with 0.08–0.16 g of lake or lake blend per liter (⅛–¼ oz/12 gal) of the product.

Troubleshooting

BEVERAGES		
Symptoms	Causes	Changes to Make
Color of reconstituted powdered beverage is too light.	Colorant dosage too low	Increase colorant dosage.
	Dry colorant not dissolving properly	Use FD&C dyes rather than lakes. Reformulate dry colorant to dissolve more rapidly.
	Insufficient mixing	Increase mixing time.
Primary colors are seen as the powdered beverage mix is added to water	Colorant flashes	Use nonflashing colorants.
Color fades during processing	Colorant sensitive to processing temperatures	Use a more heat-stable colorant. Increase colorant dosage to compensate. Adjust process to minimize exposure to high temperatures. Adjust point of addition of colorant to minimize exposure to high temperatures.
Color of final product is too light, changes in hue, or disappears	Colorant concentration too low	Use more colorant.
	Colorant sensitive to oxygen incorporated during processing	Use a more oxygen-stable colorant. Minimize oxygen incorporation during processing. Use an antioxidant in conjunction with colorant.
	Colorant-ingredient interaction	Look for negative interactions with salts, proteins, or other ingredients.
	Selective fading of one of the components of a colorant blend	Reformulate blend to remove less-stable component. See suggestions above for stabilizing colorants.
Color of final product too dark	Colorant concentration too high	Use less colorant.
Color of final product dull	Background color from other ingredients interferes with colorant	Take steps to minimize background color from other ingredients.

Symptoms	Causes	Changes to Make
Final product color fades with time on shelf	Light-induced degradation of colorant	Use a more light-stable colorant. Use FD&C lakes instead of dyes. Use an antioxidant in conjunction with colorant. Eliminate clear window on package. Use UV-absorbing film on package window.
Final product color of vitamin C-fortified beverage fades and/or turns brown with time on shelf	Ascorbic acid content of beverage causes color degradation	Use a colorant that is less susceptible to ascorbic acid-induced degradation. Use an antioxidant in conjunction with colorant.
Final product color of beverage turns brown and dull with time on shelf	Degradation of colorant	Use a more stable colorant. Where appropriate, use an antioxidant in conjunction with colorant.
Colored ring at neck of bottle	Incomplete emulsification of flavor oils and migration of color into oil phase	Increase emulsifier concentration. Use a more powerful emulsifier. Use a preblend colorant and flavor system with the correct emulsifier.
Colorant settles out	Colorant not soluble in product	Use a more soluble colorant preparation. Use a more soluble colorant.
	Colorant not soluble at product pH	Use a colorant soluble at product pH. Use a pH-stabilized version of colorant.
Can corrosion	Interaction of colorant and carbon dioxide	Use less than 50 ppm of azo colorants. Use Brilliant Blue or Fast Green. Select a can with a more resistant coating.

DAIRY PRODUCTS AND SPREADS

Symptoms	Causes	Changes to Make
Dramatic color loss or change	Microbial degradation of FD&C Red No. 40	Switch to beet, carmine, or cochineal.
Color not uniform throughout product	Insufficient mixing	Increase mixing time. Change point of addition to allow more mixing time.
	Colorant doesn't mix with product primary phase	Use a colorant carrier that will mix with product primary phase.
	Colorant present in very small quantity	Use a more dilute colorant preparation.
Specks in finished product	Precipitation or sediment in colorant preparation	Filter colorant before using.
	Colorant is not compatible with product pH	Change to colorant compatible with pH or use pH-stabilized version.

Color of final product too light, changes in hue, or disappears	Colorant concentration too low	Use more colorant.
	Colorant sensitive to processing temperatures	Use a more heat-stable colorant. Increase colorant dosage to compensate. Adjust processing conditions to minimize exposure to high temperatures. Adjust point of addition of colorant to minimize exposure to high temperatures.
	Colorant sensitive to oxygen incorporated during processing	Use a more oxygen-stable colorant. Minimize oxygen incorporation during processing. Use an antioxidant in conjunction with colorant.
	Colorant-ingredient interaction	Look for negative interactions with salts, proteins, or fortification ingredients.
	Selective fading of one component of a colorant blend	Reformulate blend to remove the less-stable component. See suggestions above for stabilizing colorants.
Color of final product too dark	Colorant concentration too high	Use less colorant.
Color of final product dull	Background color from other ingredients interferes with colorant	Take steps to minimize background color from other ingredients.
Final product color fades with time on shelf	Light-induced degradation of colorant	Use a more light-stable colorant. Use an antioxidant in conjunction with colorant. Eliminate clear window on package. Use UV-absorbing film on window.
Color of product becomes pink or salmon (pinking) with exposure to light or heat	Interaction of annatto colorant and other ingredients	Switch to β-carotene or apo-carotenal. Switch to FD&C colorants. Change ingredients to eliminate pinking.

Reference

1. Warner Jenkinson Color Service Laboratory. 1996. Natural color beverage stability study. Warner Jenkinson Co. Inc., St. Louis, MO.

CHAPTER 14

Confections

Although most confections contain some added colorant, those that are most commonly colored include candy starch jellies, candy cream centers, pan-coated candies, hard candy (boiled sweets), direct-compression tablets, oil-based (summer) coatings, hard candy wafers, and chewing gum.

Candy Starch Jellies and Candy Cream Centers

CANDY STARCH JELLIES

Candy starch jellies are made principally from sugar, water, and a modified starch. The mixture is cooked to a definite moisture content, and the hot, fluid jelly is then deposited by machine into starch molds. The size and shape of the mold depressions determine the form of the finished candy pieces.

When candy starch jellies are colored, the liquid market form of a water-soluble colorant or some water-soluble (heat-stable) natural colorant is recommended. The colorant solution should be added at the end of cooking, allowing enough time for stirring to obtain color uniformity before the batch is turned from the kettle. A concentrated colorant solution (about 59.92 g of full-strength colorant per liter [8 oz per gallon]) should be used. This prevents the moisture content of the batch from increasing significantly.

Ordinarily, the object is to produce bright, clear colors that suggest fruit flavors in the finished piece. Dull or deep shades can be caused by too much colorant, prolonged cooking, or a final product with dimensions that are too thick.

Because jellies are semitransparent, the thickness of each piece has a definite bearing on the quantity of colorant needed. Thick pieces appear more deeply colored than thin pieces from the same candy batch. However, when sugar sanding is employed as a finishing step, the shades lighten. The only way to determine the color effect that will be obtained from a given volume of a particular colorant solution is to prepare an exact sample of the finished product.

In This Chapter:

Candy Starch Jellies and Candy Cream Centers
  Candy Starch Jellies
  Candy Cream Centers

Pan-Coated Candies
  Coating Equipment
  Sugar Panning
  Exempt Colorants (Natural Colors)

Hard Candies (Boiled Sweets)

Other Products
  Direct-Compression Tablets
  Oil-Based Summer Coatings
  Hard Candy Wafers
  Gum Products

Troubleshooting

CANDY CREAM CENTERS

The cream used in making cream centers and bonbons is readily colored. The required quantity of liquid colorant is stirred into the cream while it is still warm. The artistic possibilities are great, but since delicate tints are preferred, care should be taken not to overcolor. Colorants similar to those described below are applicable.

When using natural colorants, consider the following comments and precautions.

Reds. *Cochineal extract.* Cochineal extract is orange at a pH below 4.0 and magenta red at a pH above 4.0. Although it is heat stable, it should be added at the end of heat processing, as is done with certified colorants. It is most commonly available in a water-based liquid form of 3–4% carminic acid.

Carmine. Carmine lake that has been rendered water soluble by alkali is called water-soluble carmine. This type of carmine can be incorporated from a water-based market form to yield bright, magenta red shades that are more blue red and less orange than FD&C Red No. 40. Various shades of red can be achieved by blending carmine with a yellow component such as β-carotene or annatto extract.

Beet juice color. Beet juice provides another magenta red color. Although it has poor heat stability, it can be used successfully in hot, processed candies if it is added as near the end of heat processing as possible. One of the drawbacks to beet juice is its low tinctorial strength and resulting high usage levels. Care must be taken to ensure that the final moisture level of the product is appropriate.

Anthocyanins. As described earlier in this book, anthocyanins are found in commercially available concentrated color systems from sources including red cabbage and elderberry, black currant, and grape juice concentrates (grape skin extract is permitted in the United States only in beverages). Anthocyanins are red and stable only in systems in which the pH can be 3.8 and below. Typically, use in candy starch jellies is not appropriate.

Orange shades. *Paprika oleoresin.* Although naturally oil soluble, paprika oleoresin can be rendered water dispersible with suitable emulsifiers and processing. It will not yield the bright orange shade obtained with FD&C Yellow No. 6, however. Also, one must keep in mind that an opaque orange shade is obtained with paprika oleoresin because of its oil-based nature. Stability should always be monitored when paprika oleoresin is used.

β-Carotene. β-Carotene, another naturally oil-soluble carotenoid, exhibits yellow to orange shades, depending on the product and usage level. Most commercially available water-dispersible β-carotene is available in powdered forms that do not readily disperse into a hot sugar matrix. Stock "solutions" (actually dispersions) of the powder must be prepared by dispersing it evenly in water, which can then be added at the appropriate level to the candy base and mixed well.

A liquid emulsion market form of β-carotene is the easiest for the candy maker to incorporate.

Canthaxanthin. Canthaxanthin is a synthetically made carotenoid that is available as part of a sugar matrix (beadlet or powder). It can be incorporated into a candy system after being predispersed in water. A cloudy, red-orange shade results.

Yellow. *Turmeric.* A bright lemon yellow shade can be achieved by adding a water-soluble turmeric product. Turmeric is very sensitive to degradation by light, so care must be taken when selecting the type of package.

β-Carotene. To obtain yellow colors, use low quantities (e.g., 3–6 ppm) of pure β-carotene.

Brown. Caramel colorants work well in this type of candy application. A double-strength liquid caramel colorant is easily incorporated.

Pan-Coated Candies

In pan work, colorants are applied in the coating syrup. Although water-soluble colorants are still used in some coating processes in the industry today, lake pigments in the form of dispersions are most commonly employed.

When coloring coated products with water-soluble colorants, liquid colorants are most often used. A measured volume of liquid colorant is worked uniformly into each gallon of syrup, and measured volumes of colored syrup are then added progressively to the charge in the coating pan. As a general rule, more colorant is required for coatings than for jellies or cream centers because the color of coated candy is seen on the surface rather than in depth. As many as 30–60 coats of coloring syrup are necessary to produce satisfactory shades with good color depth when water-soluble colorants are used. In addition, color development is relatively slow in this situation. Also, because the colorants are water soluble, they travel with any moisture in the pan. Therefore, it is necessary that each coat dry slowly and thoroughly. All of these factors lead to increased process time and very little margin for error.

Although opaque coatings can be achieved with a combination of colorants and titanium dioxide, the best approach for sugar-coated products is to use FD&C lake pigments, which require fewer coats on these tableted candies. Unlike water-soluble colorants, lakes color by dispersion. Since the pigment is insoluble, moisture migration is not a defining factor in the panning process. The overall procedure becomes more forgiving and allows for operator-to-operator differences. The pigment's inherent opacity promotes better coverage and quicker, more consistent color development. Since each coat of syrup made with lakes acts independently of the preceding coat, only 12–18 coats are needed to adequately color the candies.

Lakes are typically incorporated into the panning process as part of a sugar-based dispersion. This is because of the electrostatic nature of lakes, which results in loose agglomerates that require energy to break down. To achieve the maximum coloring strength of the lakes, these agglomerates must be reduced to their primary particle size.

If sufficient mixing is not available in the manufacturing process to completely disperse the pigments, specks and streaks will appear in the end product. Sugar-based dispersions of lakes are made by the color manufacturer, who incorporates the lakes with high shear. These dispersions work especially well with a coater's sugar syrup at dilutions ranging from 1:6 to 1:12.

In addition to the colorants, a typical sugar-based dispersion contains stabilizers, preservatives, and in some cases viscosity-modifying agents. To provide a stable product with maximum shelf life, factors such as shade, strength, viscosity, functionality, cost effectiveness, and international requirements must be considered when ingredients for each custom-made dispersion are selected.

COATING EQUIPMENT

Of the many types of panning equipment available today, the most common is the conventional rotational pan developed during the mid-1800s. The product tumbles inside the pan, which is mounted on a stationary arm that rotates it on a horizontal axis an average of 25–30 revolutions per minute. Many of these pans are fitted with ribs or baffles inside to prevent the tablet bed from sliding down the smooth inner walls.

During the 1960s, a new type of pan was developed. This enclosed system with a perforated tumbling pan allows for efficient airflow over the tablet bed. Inlet air temperatures and other panning parameters such as drying time, relative humidity, and spray rate can be controlled with a computer interface, greatly reducing operator involvement and increasing productivity.

SUGAR PANNING

The panning procedure consists of three steps: sealing, subcoating, and applying colorant. The coating syrup, to which colorant is added, is a simple sucrose syrup, typically with a solids content of 68–72%. The higher the solids content, the less water needs to be driven off during coating, thus lessening process time and minimizing the chance of "mottling" in the color application. A sugar shell develops with repeated applications of colored coating syrup to the tablet bed, which should be at a temperature of 35–40°C. The quantity of coating syrup added to the tablet charge should be just enough to wet the entire surface of the tablets. Too little syrup creates dry spots, resulting in a blotchy appearance. Too much syrup makes it difficult to drive off the moisture on the tablet bed, and the colorant will be literally washed off the tablets and onto the sides of the pan. Box 14-1 describes how panned candies are polished.

EXEMPT COLORANTS (NATURAL COLORS)

As U.S. manufacturers seek expanded export opportunities, natural colorants are becoming the colorants of choice because of their worldwide acceptance. In the past, sugar panning with natural colorants was erroneously thought to yield muted, unattractive coatings. Progressive technology has provided the industry with an expanded palette of shades derived from natural colorants. With consideration given to each colorant's unique properties, appropriate systems can be developed. Turmeric, β-carotene, and red and purple shades of carmine, titanium dioxide, beet juice, and red cabbage have all been successfully used in commercial panned candies in the United States.

Hard Candies (Boiled Sweets)

Batches of hard candy will not tolerate much, if any, added water after cooking. This means that water solutions of soluble colorants can be used only when the colorant is added at the cooker and the water is boiled off. Generally, water solutions are confined to items in which an entire day's vacuum-cooker run consists of the same product without a change of colorant formula. Also, it is difficult to obtain bright shades with water solutions because prolonged cooking has a dulling effect. Thus, water solutions find only limited use in hard candy.

Some companies offer premeasured, batch-size packets of specific colorants. These colorants are typically dispersed or solubilized in a hard candy-compatible matrix. The color packet is added to the top of the hot candy slab. As it "melts," the colorant is evenly dispersed in the candy.

Box 14-1. Polishing Candy Tablets

Because every confection is unique in composition and shape, many polishing methods have been developed. Tablets receive one or two "half-dose" applications of clear syrup at the end of the color-coating step to prepare them for polishing. Drying air is turned off at this stage so that the tablets dry very slowly. Once the tablets appear "velvety," they are removed from the coating pan and placed into trays or hammocks to condition them for the application of gloss.

Gloss can be achieved with one of many products, most commonly carnauba wax, beeswax, or a combination of the two. The wax can be added to the pan as a fine, dry powder or as a slurry in alcohol or mineral oil. Varieties of food-grade shellac with additives to promote shine are gaining in popularity as finishing agents. These products also form excellent moisture barriers.

Propylene glycol- or glycerin-based dispersions of lakes can also be used in hard candy applications, especially in candy canes and peppermints. The use of lakes in these systems results in improved light stability compared with that of water-soluble colorants and also gives a migration-free product.

Other Products

DIRECT-COMPRESSION TABLETS

Lakes can be dry blended (typically 0.1%) with granular forms of dextrose, sucrose, mannitol, or sorbitol to produce attractive colorants for the popular direct-compression sweet and sour candies as well as mints and lozenges. It is noteworthy that these types of tablets and confections would not have been possible without the introduction of lake pigments in 1959.

OIL-BASED SUMMER COATINGS

Hard-fat candy (or "summer") coatings, such as those used on enrobed candies, can be colored by dispersing the FD&C lakes in the coating. An oil-based dispersion works best, but the lakes can also be predispersed in vegetable oil, propylene glycol, or glycerin before use. These dispersions can then be added directly to the melted coatings in the mixer and blended for uniform and color-pleasing products. Water-soluble colorants have also been used, but they must first be dissolved in propylene glycol or glycerin with an added emulsifier. Lakes have inherently better use characteristics and greater light stability than water-soluble colorants.

HARD CANDY WAFERS

For years, water-soluble colorants have been used for coloring hard candy wafers. FD&C Red No. 3 has traditionally been used in pink, mint-flavored wafers in spite of a continuing problem with fading. However, carmine, which offers enhanced resistance to fading in sunlight, can easily be incorporated into the candy. For adequate dispersion, the pigments should be added to the batch in an arm-type mixer or an equivalent high-torque shear mixer.

GUM PRODUCTS

For good color retention in stick or ball gum, lakes provide the best colorant choice. FD&C lakes, carmine, and turmeric remain bright and attractive yet do not stain the mouth. These colorants also perform well in sugar-coated gum products and in stick gum onto which a colored design is externally added. For application on a stick gum, lakes are dispersed in an edible, film-forming agent, such as confectioner's glaze, and then applied. Lakes predispersed in propylene glycol or glycerin work very well in soft bubble gums as well.

Troubleshooting

CANDY STARCH JELLIES AND CREAM CENTERS		
Symptoms	Causes	Changes to Make
Color not uniform, specking	Dry colorant not dissolving properly	Predissolve colorant in some of process water. Use a liquid colorant preparation.
	Insufficient mixing	Increase mixing time.
	Colorant preparation too concentrated for good dispersion	Use a more dilute colorant preparation.
Color too light or fades during processing	Colorant dosage too low	Increase color dosage. Use FD&C lakes instead of dyes.
	Colorant degrades due to high process temperatures	Use a more heat-stable colorant. Increase colorant dosage to compensate. Adjust process to minimize exposure to high temperatures. Adjust point of addition of colorant to minimize exposure to high temperatures.
Color of final product changes in hue or disappears	Colorant degrades due to ingredient interactions	Reformulate to eliminate negative ingredient interactions. Use a colorant that does not interact with the other ingredients.
	Colorant sensitive to oxygen incorporated during processing	Use a more oxygen-stable colorant. Minimize oxygen incorporation during processing. Use an antioxidant in conjunction with colorant.
	Selective fading of one component of colorant blend	Reformulate blend to remove less-stable component. See suggestions above for stabilizing colorants.
Color of final product too dark	Colorant concentration too high	Use less colorant.
Color of final product dull	Background color from other ingredients interferes with colorant	Take steps to minimize background color from other ingredients.
	Partial degradation of colorant due to heat	Minimize exposure of colorant to high temperatures.
	Too much colorant	Decrease amount of colorant.
Final product color fades with time on shelf	Light-induced degradation of colorant	Use a more light-stable colorant. Use an antioxidant in conjunction with colorant. Eliminate clear window on package. Use UV-absorbing film on window.
Color of final product cloudy or hazy	Colorant preparation includes lakes, colorant particles, or color emulsion droplets	Reformulate colorant preparation to use soluble colorants or produce a finer colorant emulsion.

CHAPTER 15

Special Topics

Handling Colorants and Maintaining Color Quality

HANDLING WATER-SOLUBLE COLORANTS

When a shipment of a food ingredient arrives at a production facility, it is customarily inspected to determine whether or not it meets specifications. This may not always be necessary, especially in the case of primary certified food colors. Not only does the colorant manufacturer generally analyze every batch of colorant with ultramodern instrumentation, but, in the United States, FDA also conducts an independent analysis and certifies that every batch meets the U.S. purity specification (hence the term "certified food colors"). The FDA certification number is included on the label of every container and on every invoice of material certified in the United States.

There are two methods for checking the strength of a particular shipment of certified food color: titration with titanium trichloride (which is not recommended for blends) and spectrophotometric analysis. Details of these methods are beyond the scope of this book but are available from colorant suppliers.

A more common situation is the need to determine whether a given batch of blend will provide the desired color in the finished product. To make this determination, a random sample is taken, and the visible spectrum of the sample is generated and compared with the spectrum of the standard. If the two match, the shipment is accepted. As a further check, the colorant is usually inspected visually after it has been prepared for the spectrophotometric evaluation. It is important that the colorant be checked after it has been dissolved rather than examining the dry colorant in its container so that small, unimportant changes in physical characteristics that alter the colorant's outward appearance do not lead to erroneous rejection of a shipment.

PREPARING WATER-SOLUBLE COLORANTS FOR USE

Powders and granules can be prepared for use by dissolving them in water. Before dissolving colorant, check the solubility information provided by the manufacturer. For most colorants, 15–32 g are used per liter (2–4 oz/gal), depending upon the shade desired. FD&C Blue

> **In This Chapter:**
> Handling Colorants and Maintaining Color Quality
>   Handling Water-Soluble Colorants
>   Preparing Water-Soluble Colorants for Use
>
> Factors Affecting Colorant Quality

No. 2 is not suitable for stock solutions because of its low solubility and lack of stability.

To prepare 1 gal, weigh the desired amount of colorant and add it to three quarts of warm water. Agitate the water until all the colorant has been dissolved, and then add water to make 1 gal.

If the color solution will be kept overnight or longer, it must be refrigerated or a preservative must be added. For colorants containing FD&C Red No. 3, the colorant solution can be preserved with 15% (wt/wt) propylene glycol. Solutions of the other colorants can be preserved by adding 4.5 ml of a 23% sodium benzoate solution per liter plus either 5.32 ml of 61% phosphoric acid solution or 5.32 ml of 50% citric acid solution per liter.

The solution can be filtered through fast filter paper in a glass funnel to remove sediment.

The following should be considered when colorant solutions are prepared:

1. Distilled water, demineralized water, or zeolite-softened water should be used. If hard water must be used, remember that calcium and magnesium ions can form sediment consisting of insoluble salts, which will appear as a sludge in the bottom of the container.

2. The vessels and accessories used in the dissolving process must not react with the colorants. Stainless steel containers are excellent. For small-scale work, stainless steel beakers, pails, dippers, and stirrers are easy to obtain. For larger operations, stainless steel agitators, valves, and fittings are recommended. Glass or glass-lined equipment, monel, inconel, and plastic are also suitable.

3. The dry colorant should always be weighed, not measured with a scoop, cup, or other container. The density of a dry colorant varies from batch to batch for a number of reasons. These variations have no effect on color purity or quality; however, consistent results cannot be obtained if dry colorants are measured volumetrically.

4. The colorant must be completely dissolved. Specking, spotting, weak colors, and other unsatisfactory outcomes can sometimes result from inadequate stirring. The solution is ready only after the liquid is clear, bright, and unclouded and no undissolved colorant is left sticking to the sides or bottom of the vessel.

5. If preservatives are used, they must be added in the correct order. Sodium benzoate is added first and then the acid. If the order is reversed, benzoic acid will precipitate, and the colorant solution will become cloudy and not be preserved.

The easiest and most reliable way to avoid problems and to be certain that the colorant solution will produce the desired results in the finished product is to purchase the liquid market form.

Factors Affecting Colorant Quality

To achieve the best results, water-soluble colorants should be protected from oxidizing agents, reducing agents, contamination with

> Beginners often make the mistake of adding more colorant in an attempt to create hues that are brighter and more intense. In fact, too much color makes foods darker rather than more richly colored and results in an unappetizing product. Therefore, as little color as possible should be added to obtain the final shade.

microorganisms, strong acids and alkalis, excessive heat, and intense light.

Oxidizing agents. Oxidizing agents such as ozone, chlorine, and hypochlorites quickly decolor solutions of soluble colorants. Therefore, dry colorants, color solutions, and colored products should be kept away from bleaching agents.

Reducing agents. Reducing agents such as sulfur dioxide, invert sugars, some flavors, metallic ions, and ascorbic acid can induce color changes, the rate of which depends upon the individual colorant involved. Information about the stability of various primary soluble colorants to reducing agents can often be obtained from the manufacturer.

Metals, especially aluminum, zinc, tin, and iron, also act as reducing agents and should be eliminated from handling and storage vessels, pipe lines, fittings, product containers, and caps. Copper should not be used when dissolving, handling, or storing colorant solutions. However, copper kettles, such as those used in candy manufacturing, do not cause a problem when the colorant is added to essentially neutral products and exposure to the metal is not prolonged.

Contamination with microorganisms. Microorganisms, particularly molds and reducing bacteria, induce severe color fading. However, this property of soluble colorants is considered an advantage because when color fades as a result of such activity, it is an indication that the food is unfit for consumption.

Acids and alkalis. The effects of acids and alkalis on color stability depend on the other ingredients present. For example, the activity of fading agents, especially metals, can be greatly increased by either very high or very low pH. Soluble colorants are generally stable in the presence of food acids such as citric and malic acid, but one must be careful about solubility. For instance, FD&C Red No. 3 is insoluble in acid media, so normally it should not be used in food systems with a pH of less than 5.0.

Heat. None of the soluble colorants, except FD&C Red No. 3, is stable under conditions of prolonged high temperatures. In cooked foods, it is safest to add the colorant as late in the cooking process as possible, preferably after the batch has cooled somewhat. Also, the activity of oxidizing and reducing agents is enhanced by elevated temperatures, so these materials must be eliminated whenever possible before colorant is added.

Light. Light is ordinarily not a major problem, except if a product is exposed to prolonged sunlight, for example, in a display window. However, light fastness does vary considerably among the primary soluble colorants. FD&C Red No. 40 and Yellow No. 5 have moderate light stability, while FD&C Blue No. 2 and Red No. 3 have poor light stability. Therefore, it is best to minimize exposure to direct sunlight, particularly when blends are involved. Reactions of FD&C colorants under a variety of conditions are discussed in Chapter 5.

Shelf life. Synthetic soluble colorants are quite rugged compared with the "natural" or "nature-identical" food colorants. Nevertheless, unnecessary exposure to heat, light, and reducing agents should be avoided. Containers should be kept tightly sealed when the colorant is not being withdrawn, and synthetic food colorants should be stored in a relatively cool place. Because shelf life is not a problem for soluble colorants when ordinary precautions are taken, several months' inventory can be maintained without fear of degradation. However, when inventory levels are determined, personal risk preferences, cost, and colorant availability (including availability of intermediates, vagaries of the FDA certification process, and marketplace demand) should also be considered.

CHAPTER 16

Future Prospects

The preference for natural colorants over synthetics started with the green movement of the 1960s and shows no sign of decreasing. This may result from a perceived uneasiness with the safety of the FD&C colorants on the part of the consumer, but another factor, perhaps more important, is that most governments allow more flexibility and leniency in the use of natural colorants.

With the exception of FD&C Red No. 40, which was introduced in 1971, and the recent petitioning for D&C Red No. 28 for food use, there is little enthusiasm for the development of new FD&C colorants. An important factor is the cost of toxicological studies involved in the approval of a new colorant.

Considerable research interest has been shown in the biotechnology of food colorants (1), particularly in the production of existing colorants with a predictable market. Products of biotechnology will have to compete with traditional agricultural sources of supply in terms of both economics and quality. One advantage for biotechnology is the assurance of a dependable supply not subject to the vagaries of climate and politics.

Recent research on biotechnology may be divided into three areas: plant cell tissue culture, microbial fermentation, and enzyme and gene manipulation techniques. Plant tissue culture techniques are well established for commercial production of plants for the horticultural trade. The single, large-scale commercial success in the production of a colorant involves the colorant shikonen. Anthocyanin production from carrots and grapes, betanin from beets, bixin from annatto, and crocin from crocus have all received research attention. Microbial fermentation combined with genetic selection has been successful with β-carotene from *Dunaliella salina* and *Blakeslea trispora*, astaxanthin from *Phaffia rhodozyma* and *Haematococcus* spp., lutein from *Chlorella pyrenoidosa*, zeaxanthin from *Flavobacterium* spp., canthaxanthin from *Cantharellus cinnabarinus*, and phycocyanin from *Spirulina*. The microbial growth and fermentation procedures may be combined with modification of the growth environment to optimize enzyme action and colorant production, for example, in DNA transfer techniques. The transfer of the gene from the insect *Dactylopius coccus*, which produces cochineal, to a yeast will extend the potential for colorant production. These approaches may have a strong impact on the traditional sources of colorants.

Dozens of new pigments are reported in the literature every year (2), so there is no shortage of ideas. Success will lie in the ability to commercialize new sources and new pigments for food colorants, including taking them through the regulatory process.

References

1. Francis, F. J. 1996. Less common natural colorants. In: *Natural Food Colorants*, 2nd ed. G. A. F. Hendry and J. D. Houghton, Eds. Blackie Publishers, Glasgow, Scotland.
2. O'Callaghan, M. C. 1996. Biotechnology in natural food colours. In: *Natural Food Colorants*, 2nd ed. G. A. F. Hendry and J. D. Houghton, Eds. Blackie Publishers, Glasgow, Scotland.

Appendix

Description of Colorants

The colorants discussed in this book are listed alphabetically in this appendix. The chart shows hue, legal status in the United States as a food colorant, source, and section in the *U.S. Code of Federal Regulations*, as well as the European number, CAS Registry code number, and CI number and name when available. The name of each colorant is repeated on the right page of the spread as an aid in reading across the columns.

Description of Colorants

Colorant Name	Hue	U.S. Legal Status (as a Colorant for Foods)	Source
Algae extract	Brown	Not allowed	Extraction of seaweed
Algae extract	Red	Not allowed	Extraction of seaweed
Algae meal	Yellow	Exempt. For chicken feed only	Dry algae cells
Alkannet	Red	Not allowed	Roots of plants (*Alkanna* or *Alchusa* spp.)
Alumina	White	Not allowed	Aluminum hydroxide
Aluminum powder	White	Not allowed	Powdered aluminum
Annatto	Reds, yellows	Exempt	Tropical shrub, *Bixa orellana*
β-Apo-8′-carotenal	Orange	Exempt, with restrictions	Synthetic
Astaxanthin	Red	Allowed only in fish food	By-product of lobster and shrimp processing. Also synthetic and from algae
Beet juice concentrate	Red to yellow	Allowed, as vegetable juice	Beet, *Beta vulgaris*
Beets, dehydrated (powder)	Red	Exempt	Beet, *Beta vulgaris*
Betanin	Red	Not allowed	Pigment of beets
Cabbage, red	Red	Exempt. Permitted as an ingredient	Red cabbage
Cacao	Brown	Not allowed	Cacao plant (*Theobroma cacao*)
Calcium carbonate	White	Permitted as functional additive	Calcium carbonate, precipitated from solution
Canthaxanthin	Red	Exempt. For chicken feed only	Synthetic
Caramel	Brown	Exempt	Heating of sugars
Carbon Black	Black	Not allowed	Combustion of vegetable material to residual carbon
Carmine	Red	Exempt	Lake of cochineal
Carmoisine (azorubine)	Red	Not allowed	Napthionic acid and derivatives
β-Carotene	Orange, yellow	Exempt	Synthetic. Also from molds, microalgae (e.g., *Dunaliella*)
zeta-Carotene	Yellow	Not allowed	Tomatoes
Carrot oil	Yellow	Exempt	Carrots
Carthamin	Yellow to red	Not allowed	Safflower flowers (*Carthemus tinctorius*)
Citrus peel	Orange	Not allowed	Citrus peel
Citrus Red No. 2	Red	Certified for citrus orange skins only	Synthetic
Cochineal extract	Red	Exempt	Dried bodies of female cochineal insects (*Dactylopius coccus*)

Colorant Name	21 CFR[a]	EU[a] No.	CAS[a] Reg. No.	CI[a] No. and Name
Algae extract			977026 928	
Algae extract			977090 042	
Algae meal	73.275			
Alkannet				
Alumina		E 173		
Aluminum powder		E 173		77000, Pigment Metal 1
Annatto	73.30	E 160b		75120, Natural Orange 4
β-Apo-8′-carotenal	73.90	E 160a		
Astaxanthin	73.35			
Beet juice concentrate	73.260			
Beets, dehydrated (powder)	73.40	E 162		
Betanin		E 162	7659-95-2	
Cabbage, red	73.260	E 163		
Cacao				
Calcium carbonate	73.1070	E 170	471-34-1	77220, Pigment White 18
Canthaxanthin	73.75	E 161g		
Caramel	73.85	E 150		Natural Brown 10
Carbon Black				
Carmine	73.100	E 120	1390-65-4	
Carmoisine (azorubine)				14720
β-Carotene	73.95			
zeta-Carotene				
Carrot oil	73.300			
Carthamin				75140, Natural Red 26
Citrus peel				
Citrus Red No. 2	74.302			
Cochineal extract	73.100	E 120		75470, Natural Red 4

Description of Colorants, continued

Colorant Name	Hue	U.S. Legal Status (as a Colorant for Foods)	Source
Copper chlorophyllin	Green	Not allowed in foods	Processed from green leaves
Copper pheophytin	Gray-green	Not allowed in foods	Processed from green leaves
Corn endosperm oil	Reddish brown	Exempt. For chicken feed only	Extraction of corn gluten
Cottonseed flour	Yellow	Exempt, with restrictions	Processed cottonseed kernels
Cottonseed kernels	Brown	Not allowed	Processed cottonseed kernels
D&C Red No. 28 (Phloxene B)	Red	Not allowed but under consideration	Synthetic
D&C Yellow No. 10 (Quinoline Yellow)	Yellow	Not allowed	Monosulfomonosodium salt of disulfonic acid
FD&C Blue No. 1 (Brilliant Blue)	Greenish blue	Certified	Synthetic
FD&C Blue No. 2 (Indigotine)	Deep blue	Certified	Synthetic
FD&C Green No. 3 (Fast Green)	Bluish green	Certified	Synthetic
FD&C Red No. 3 (Erythrosine)	Bluish red	Certified	Synthetic
FD&C Red No. 40 (Allura Red)	Yellowish red		Synthetic
FD&C Yellow No. 5 (Tartrazine)	Lemon yellow	Certified	Synthetic
FD&C Yellow No. 6 (Sunset Yellow)	Reddish yellow	Certified	Synthetic
Ferrous gluconate	Brown	Exempt. For ripe olives only	Synthetic
Fruit juice concentrates	Various	Exempt	Various
Gold	Gold	Not allowed	Gold flakes
Grape Color Extract	Red to blue, depending on pH	Exempt. For nonbeverage foods only	From the fruit
Grape Skin Extract	Red to blue, depending on pH	Exempt. For beverages only	By-product of wine and juice industry
Haem colorants	Red	Not allowed	Blood
Iridoid pigments	Green, yellow, red, and blue	Not allowed	Gardenia fruits (*Gardenia jasminoides*)
Iron oxide	Black, brown, red, yellow (ochre)	Exempt. For pet foods only	Synthetic from ferrous sulfate
Kermes	Red	Not allowed	From insects (*Kermes* or *Kermococcus* spp.)

Colorant Name	21 CFR[a]	EU[a] No.	CAS[a] Reg. No.	CI[a] No. and Name
Copper chlorophyllin				
Copper pheophytin				
Corn endosperm oil	73.315		977010 506	
Cottonseed flour	73.140		977050 546, 977043 778	
Cottonseed kernels			997043 778, 997043 78	
D&C Red No. 28 (Phloxene B)				45410
D&C Yellow No. 10 (Quinoline Yellow)		E 104		
FD&C Blue No. 1 (Brilliant Blue)	74.101			42090
FD&C Blue No. 2 (Indigotine)	74.102			73015
FD&C Green No. 3 (Fast Green)	74.203			42053
FD&C Red No. 3 (Erythrosine)	74.303			43430
FD&C Red No. 40 (Allura Red)	74.340			16035
FD&C Yellow No. 5 (Tartrazine)	74.705			19140
FD&C Yellow No. 6 (Sunset Yellow)	74.706			15985
Ferrous gluconate	73.160			
Fruit juice concentrates	73.250			
Gold				
Grape Color Extract	73.169	E 163		
Grape Skin Extract	73.170	E 163		
Haem colorants				
Iridoid pigments				
Iron oxide	73.200	E 172		Pigment Black 11; 77499, Pigment Browns 6 and 7; 77491, Pigment Reds 101 and 102; 77492, Pigment Yellows 42 and 43
Kermes				

Description of Colorants, continued

Colorant Name	Hue	U.S. Legal Status (as a Colorant for Foods)	Source
Lac	Red	Not allowed	From insect (*Laccifer lacca*)
Lycopene	Red	Not allowed	From red tomatoes
Monascus colorants	Yellow, red, or purple	Not allowed	From a fungus (*Monascus* spp.)
Octopus and squid inks	Black	Not allowed	Octopus and squid inks
Orange B	Orange	Certified for surfaces of wieners and frankfurters only	Synthetic
Palm oil	Red, orange	Not allowed	From palm tree *Elaeis guineensis*
Paprika	Deep red	Exempt	Pepper pods, *Capsicum annum*
Paprika oleoresin	Orange-red	Exempt	Extract from *C. annum*
Phycobilin colorants	Red or blue	Not allowed	Algae, various species
Riboflavin	Greenish yellow	Exempt	Synthetic
Saffron	Yellow	Exempt	Crocus bulb, *Crocus sativus*
Silicon dioxide	White	Not allowed. Permitted as functional additive	Synthetic, from silica
Silver	Silver	Not allowed. Permitted as coating on food wrappers	Silver nitrate reacted with ferrous sulfate
Tagetes	Yellow to orange	Exempt. For chicken feed only	Flower petals of Axtec marigold (*Tagetes erecta*)
Talc	White	Not allowed. Permitted as a release agent in baking	Magnesium silicate
Tea	Brown	Not allowed	Extracts of tea (*Thea sinensis*)
Titanium dioxide	White	Exempt, allowed up to 1% of weight of final product	A mineral, ilmenite ($FeTiO_2$)
Turmeric	Bright yellow to greenish yellow	Exempt	Dried, ground rhizomes of herb (*Curcuma longa*)
Turmeric oleoresin	Bright yellow to greenish yellow	Exempt	Extraction of *Curcuma longa* rhizomes
Ultramarine Blue	Blue	Exempt. Salt for animal feed only	Synthetic, from minerals
Ultramarine Green	Green	Not allowed	Synthetic, from minerals
Ultramarine Violet and Red	Violet and red	Not allowed	Synthetic, from minerals
Vegetable juice	Various	Exempt	Various
Xanthophyll paste	Green	Not allowed	Extracts of alfalfa, nettles, or broccoli
Zinc oxide	White or yellowish white	Permitted in food wrappers and as a nutritional supplement	Zinc

[a] CFR = *Code of Federal Regulations*, EU = European Union, CAS = Chemical Abstract Service, CI = color index system of the American Association of Textile Chemists and Colorists.

Colorant Name	21 CFR[a]	EU[a] No.	CAS[a] Reg. No.	CI[a] No. and Name
Lac				
Lycopene				
Monascus colorants				
Octopus and squid inks				
Orange B	74.250			
Palm oil				
Paprika	73.340	E 160c		
Paprika oleoresin	73.345	E 160c		
Phycobilin colorants				
Riboflavin	73.450	E 101		
Saffron	73.500			75100, Natural Yellow 6
Silicon dioxide	175.300			
Silver		E 174		
Tagetes	73.295			
Talc	182.90		14807-96-6	77018, Pigment White
Tea				
Titanium dioxide	73.575	E 171	1 3463-67-7	77891, Pigment White
Turmeric	73.600	E 100 (curcumin)	458-37-7	75300, Natural Yellow 3
Turmeric oleoresin	73.615			
Ultramarine Blue	73.50	E 180		77007
Ultramarine Green				77013
Ultramarine Violet and Red				
Vegetable juice	73.260			
Xanthophyll paste				
Zinc oxide	182.8991		1314-13-2	77947, Pigment White

Glossary

Additive colorimetry—Analysis of colors produced by the combination of three or more primary colors (e.g., red, green, and blue) or by the combination of spectral colors.

ADI—The acceptable daily intake, the amount of a specific food additive thought to be the maximum level that should be ingested on a daily basis , as determined by experts such as those at WHO/FAO. Usually includes a 100-fold safety factor, but the safety factor may range from 10 to 5,000. Usually expressed as milligrams of the test substance per kilogram of body weight per day, based on a hypothetical average person weighing 70 kg.

Aglycone—The component in a flavonoid pigment (one of two or more compounds in the same molecule) that contributes most of the color.

ASTA—American Spice Trade Association

CAS code number—A number assigned to a colorant by the Chemical Abstract Service Registry.

Certified colorants—Colorants required by law to be certified by the FDA, e.g., FD&C Red No. 40.

CFR—The *Code of Federal Regulations*, the compilation of all U.S. laws.

Chroma—The coordinate in the Munsell System that corresponds to the color intensity scale.

Chromophore—A general term for the portion of a molecule that contributes most of the color.

CI number—The color index number assigned by the American Association of Textile Chemists and Colorists.

Color blindness—The inability to distinguish between chromatic colors.

Color coordinates—Three coordinates within a color solid that locate a specific color in three-dimensional space.

Color solid—A three-dimensional solid within which every point represents a specific color. The solid can be characterized either mathematically or physically.

Colorant, color—A *colorant* is a pigment used to color a product, as distinct from a *color*, which is used in this book in the physical sense, to indicate a hue such as red, green, blue, etc.

Cones—Conical receptors in the eye that respond to color.

D&C colors—Colors permitted by FDA for use in drugs and cosmetics.

Daily intake—Term used by the FDA to estimate the intake of a given compound from all sources.

De minimis—A concept used by regulatory agencies to describe a substance with a negligible risk.

Delaney Clause—A clause in the 1958 Food Additive Amendment forbidding the use of a substance if, after appropriate tests, any part of it was shown to cause cancer in humans or animals.

Dispersion—The suspension of very small particles in a liquid media. If the particles are small enough, the suspension is relatively stable.

E number—The number assigned to a colorant in Europe.

Enocyanin—A general term for a colorant made from grape skins or other by-products of the wine or grape juice industry.

Exempt colorants—A group of 26 colorants not required to be certified.

FD&C colorants—Colorants certified by FDA as safe for use in food, drugs, and cosmetics. Also called certified colorants.

Flavonoid—A general term for a very large group of yellow to red pigments found in many plants.

Geosmin—A chemical found in red beets that contributes most of the beet flavor.

GRAS—A Food and Drug Administration regulatory status meaning Generally Recognized as Safe.

Hue—Gradation of color. Also the coordinate in the Munsell System that corresponds to the red-green-blue scale.

Hue index—A value describing the yellow to red hue of a colorant formulation.

IUB-IUPAC—International Union of Biochemistry–International Union of Physics and Chemistry.

JECFA—The Joint Expert Committee on Food Additives of the Food and Agriculture Organization and World Health Organization.

Lake—A water-soluble colorant precipitated on a base of aluminum or calcium salts.

Monascolins—A group of microbial metabolites produced in association with the production of *Monascus* pigments.

Munsell Book of Color—A collection of color chips organized in a three-dimensional solid.

Natural colorant—A regulatory term used in other parts of the world, but not in the United States, to denote a colorant that exists in nature. The exact definition varies by country.

Pigment—A colored chemical compound.

Primary colors—Any three colors that, when mixed in suitable proportions, produce any color. Primary colors may work by subtraction of light (e.g, red, blue, and yellow) or by addition of light (e.g., red, green and blue).

Reference toxicity dose (rTD)—Term used by the U.S. Environmental Protection Agency rather than ADI.

Rods—Long, slender bodies in the eye that respond to light and dark.

Spectral color—A wavelength within the visible spectrum.

Standard of identity—A legal standard, maintained by the FDA, that defines a food's minimum quality, required and permitted ingredients, and processing requirements, if any. It applies to a limited number of staple foods.

Subtractive colorimetry—Analysis of colors produced by subtracting portions of the visible spectrum from white light.

Synthetic colorant—A colorant that does not occur in nature.

Tinctorial power—A value describing the ability of a colorant to color a product.

Tristimulus—Describing something that reacts to the three values (red, green, and blue) that make up color.

Value—The coordinate in the Munsell System that corresponds to the lightness-darkness scale.

Visible spectrum—The wavelengths of the spectrum (usually defined as 380–680 nm) that can react with the rods and cones in the eye to send a signal to the brain.

Index

a/b value, 18
Absorption, of colorant on media surface, 36
Acceptable daily intake (ADI), 24, 25–26
"Acid-proof" forms of colorants, 109, 110, 113
Additives, color
  certified, *see* Certified colorants
  exempt, *see* Exempt colorants
  generally recognized as safe (GRAS), 31
  purposes, 4–5, 89
  regulations, summary, 30
ADI, *see* Acceptable daily intake
Adulteration of foods, history, 1–2
Age, and reaction to colors, 4
Alchusa tinctoria, 75
Alfalfa, source of chlorophylls, 69
Algae and microalgae
  and carotenoids, 43, 50–51, 68
  and chlorophylls, 67
  extracts, 94, 132–133
  meal, 28, 94, 132–133
  and phycobilins, 70–71, 72
Alkannet, 76, 132–133
Alkannet tinctoria, 76
Allophycocyanins, 70, 71
Allura Red, *see* FD&C Red No. 40
Alumina, 93, 132–133
Aluminum powder, 93, 132–133
Amanita muscaria, 63
American Association of Textile Chemists, 27
American Spice Trade Association, 47
Amino acids, as colorants source, 88
Animal feeds, 50, 91, 94. *See also* Pet foods *and* Fish foods
Ankaflavin, 80
Annatto, 44, 46, 132–133
  in baked goods and cereals, 99, 100
  extract, 28, 99, 100, 109
  in pet foods, 103
Antheraxanthin, 50
Anthocyanidin, 55
Anthocyanin-3-glucoside adducts, 59
Anthocyanins, 100, 101–102
  acylated, 56, 58–59
  applications, commercial, 59–61
  chemical composition, 55–57
  commercial preparation, 57–59
  co-pigmentation effect, 58
  degradation, 61
  health benefits, 61
  stability, 58, 118
  synthetic compounds, 59
Anthraquinone colorants, 72–76
Antioxidants
  health benefits, 52, 61
  to minimize degradation, 46, 48, 51
Aphanotheca nidulans, 72
β-Apo-8′-carotenal, 28, 50, 51, 101, 132–133
β-Apo-8′-carotenoic acid, 51
Artificial colorants, 23. *See also* Certified colorants
Aspergillus japonicus, 88
Astaxanthin, 43, 50, 51, 129, 132–133
Azorubine, *see* Carmoisine
Aztec marigold, *see* Tagetes

Bacillus subtilis, 88
Baked goods, colorants in, 97–102, 103–104
Bases, for lakes, 40
Beet juice, 30, 64, 100, 103, 112, 113, 118
  concentrate, 110–111, 132–133
Beet powder, 64, 111, 132–133
Beets, 61, 64
  dehydrated, 28, 64
  extracts, 63, 65
Beta vulgaris, 61, 63
Betacyanins, 62, 63
Betalains, 62–65
Betanidin, 62
Betanin, 62, 63, 64, 65, 132–133
Betaxanthins, 62, 63
Beverages, colorants in, 107–112, 114–115
  carbonated, 107, 108
  dry mixes, 64, 107, 108, 111
Bilirubin, 70

Biliverdin, 70
Bisdemethoxycurcumin, 77
Bixa orellana, 44
Bixin, 43, 44, 46, 109
Blair Process, 67
Blakeslea trispora, 50, 129
Blending, of colorants, 20–22, 36, 38, 101
Blood, colorant source, 70
Blueberries, 101
Brilliant Blue, *see* FD&C Blue No. 1
Brown polyphenols, 86
Bureau of Chemistry, and regulation of colorants, 2, 27

C_{17} yellow pigment, 44
Cabbage, *see* Red cabbage
Cacao, 86, 132–133
Calcium carbonate, 92, 132–133
Candy, colorants in
  cream centers, 118–119, 123
  hard, 121–122
  miscellaneous, 122
  pan-coated, 119–121
  starch jellies, 117, 123
Canthaxanthin, 28, 44, 51, 129, 132–133
  applications, 101, 112, 119
Capsanthin, 43, 47
Capsicum species, 46, 47
Capsorubin, 43, 47
Caramel, 28, 132–133
  applications, 85, 99, 112, 119
  chemistry, 83
  classes of, 84
  preparation, commercial, 83
  reactants, 83, 84
Carbon Black, 90, 132–133
Carcinogens, testing for, 24
Carmine, 28, 74, 75, 100, 103, 132–133
  applications, 75, 109, 112, 113, 118
  lakes, 74, 75
Carminic acid, 73, 74–75, 118
Carmoisine, 36, 132–133
α-Carotene, 49, 50
β-Carotene, 26, 28, 101, 129, 132–133

applications, 109, 118–119, 122
properties, 43, 44, 52, 103
sources, 46, 48, 49, 50, 51
γ-Carotene, 50, 132–133
zeta-Carotene, 44
Carotenoid colorants, 43–52, 87
extracts, miscellaneous, 49–51
health aspects, 52
sources, 43, 51–52
Carrot oil, 28, 132–133
Carthamic, *see* Carthamin
Carthamin, 78–79, 132–133
Carthemone, *see* Carthamin
Carthemus, *see* Carthamin
Carthemus tinctorius, 78
CAS, *see* Chemical Abstract Service Registry
Center for Food Safety and Applied Nutrition (CFSAN), 27
Cereals, colorants used in, 97–102, 103–104
Ceroalbolinic acid, 75
Certification, of color additives
history, 27, 33
regulations, 30
Certified colorants
laws about, 27, 30
list of those currently approved, 36
CFR, *see U.S. Code of Federal Regulations*
Chemical Abstract Service (CAS) Registry, 27
Chlorophyll, 68, 69, 71
Chlorophyll colorants, 67–69
Chlorophyllide, 67, 68
Chlorophyllin, copper, 69
Chroma (Munsell), 16, 17, 19
CI, *see* Color index system
CI Natural Red 4, 73
CI Natural Red 26, 78
CI Natural Yellow 3, 77
CI Natural Yellow 6, 46
CI Pigment Black 11, 91
CI Pigment Browns 6 and 7, 91
CI Pigment Reds 101 and 102, 91
CI Pigment White, 89, 91
CI Pigment White 18, 92
CI Pigment Yellows 42 and 43, 91
CIE XYZ, *see* Measurement of color, CIE XYZ system
CIELAB, *see* Measurement of color, L* a* b* system
CIELCH, *see* Measurement of color, L* C* H* system
Citranaxanthin, 51
Citroxanthin, 50
Citrus peel extracts, 49–50, 132–133
Citrus Red No. 2, 36, 132–133
CMC color tolerancing system, 20
Coal tar dyes, 2, 3, 33, 34
Coatings, fat-based

bakery, 97, 98
candy, 119–121, 122
ice cream, 113
Cochineal, 74, 129
applications, 75, 109, 112, 113, 118
extract, 28, 73, 100, 132–133
Codex Alimentarius Commission, 26
Color, physiological basis, 7–8
Color Additive Amendment (1960), 27, 29, 34
Color blindness, 8
Color difference computer (CODIC), 20
Color index (CI) system, 27
Color mixture computer (COMIC), 21
Color solids, for color measurement, 8, 11, 15, 17
Color systems, 11, 15–17
Color tolerances, 18–20
Colorants exempt from certification, *see* Exempt colorants
Colorimeters, specialized, 12, 18
Colorimetry
additive, 20
sample presentation, 16
subtractive, 20, 21
tristiumlus, 11–12
Colors Directive (European Union), 26
Commission Internationale d'Eclairage (CIE), 10
Concentration, of colorants in solution, 98–99
Cones, in human eye, 7, 8
Confections, colorants in, 117–123
Copper chlorophyllin, 69, 134–135
Copper pheophytin, 69, 134–135
Corn endosperm oil, 28, 94, 134–135
Corrosion, of beverage cans, role of colorants, 108
Cottonseed flour, 28, 94, 134–135
Cottonseed kernels, 94–95, 134–135
Crocetin, 43, 46
Crocin, 46, 87
Crocus sativus, 46
Cryptoxanthin, 48, 50
Curcurma, *see* Turmeric
Curcuma longa, 77
Curcumin, 77, 78
Cyanidin, 56
Cyanidin-3-rhamnoglucoside, 60
Cyanidium caldarium, 72

D&C colorants, 27, 30, 33
D&C Red No. 28, 34, 129, 134–135
D&C Yellow No. 10, 36, 134–135
Dactylopius coccus, 73, 74, 129
Dairy products and spreads, colorants in, 112–114, 115–116
2-Decarboxybetanidin, 62
Delaney Clause, 29, 31, 34
Delphinidin, 56

Demethoxycurcumin, 77, 78
de minimus concept, 31
Deoxyerythrolaccin, 75
Development of new colorants, 32, 129–130
Diluants, for lakes, 40
Dispersion of color, 41
Dunaliella species, 50–51, 129

E number, 27
Elaeis guineensis, 49
Elderberry juice, 111
Enocianina, 28, 111
EPA, *see* U.S. Environmental Protection Agency
Erythrolaccin, 75
Erythrosine, *see* FD&C Red. No. 3
European Union, regulation of colorants, 26–27
Exempt colorants, 28, 103, 121
labeling requirements, 30
laws about, 27, 30
External D&C colors, 30, 33
Extraction of pigments from grapes, 57–58
Eye, human
and color stimuli, 7–8
sensitivity to color, 19

Fading, of color, 108, 109
Fast green, *see* FD&C Green No. 3
Fats, addition of colorants to, 97–98, 113
FD&C Blue No. 1, 36, 100, 101, 102, 108, 134–135
FD&C Blue No. 2, 34, 36, 100, 101, 102, 126, 127, 134–135
FD&C colorants
in baked goods and cereals, 97–99
in beverages, 107–108
history, 33–34, 36
labeling requirements, 30
laws about, 27, 30
naming, 27, 33
stability, 36, 38
FD&C Green No. 3, 36, 108, 134–135
FD&C Orange Nos. 1 and 2 (delisted), 34
FD&C Red No. 3, 34, 36, 100, 112, 113, 122, 126, 127, 134–135
FD&C Red No. 32 (delisted), 34
FD&C Red No. 40, 36, 41, 100, 103, 113, 127, 129, 134–135
FD&C Yellow Nos. 1–4 (delisted), 34
FD&C Yellow No. 5, 36, 78, 85, 101, 102, 112, 127, 134–135
FD&C Yellow No. 6, 34, 36, 101, 112, 118, 134–135
FDA, *see* U.S. Food and Drug Administration

Ferrous gluconate, 28, 134–135
Fillings, cookie, 97, 98
Fish food, colorants for, 50
Flashing, of colorants, 102, 104
Flavonoid compounds, 55, 56, 78, 87
Flavonol, 56
Flavor of foods, association with color, 3–4
Food Additives Amendment (1958), 31
Food and Drug Act (1906), 27, 33
Food, Drug and Cosmetic Act (1938), 27, 31, 33, 34
Food inspection decisions (FIDs), 2, 3, 33
Food laws, in the United States, 27–32
Food Quality Protection Act (1996), 31
Food Safety Council, 29
Fruit drinks, 60
Fruit extracts, 93
Fruit juice, 28, 111, 134–135
Fuchsine, 2
Fucoxanthin, 43
Fungus, source of colorants, 78, 79
Future prospects, 129–130

Gardenia, source of colorants, 87–88
Gardenia species, 87
Generally recognized as safe(GRAS), requirement for colorants, 31, 40
Geniposide, 88
Geosmin, 64
Gliopeltis furcata, 94
Gold, 93, 134–135
Grape Color Extract, 28, 30, 134–135
Grape juice, 111
Grape Skin Extract, 28, 30, 57, 102, 111, 134–135
Grapes, 55, 56, 93
Gravy, colorants in, 102
Gums, 122

Haem colorants, 69–70, 134–135
Haematococcus species, 51, 129
Hemoglobin, 70, 71
Hesse, Bernard, 33, 34
History of colorants, 1
Hue index, 84
Hue (Munsell), 16, 19
Hunter Rd value, 18

Icings, 98
Incidental colorants, 89
Indicaxanthin, 62
Indigotine, *see* FD&C Blue No. 2
Influence of color
  on acceptability of food, 4
  on food flavor, 3–4
  on perception of sweetness, 4
Inks, for foods, 40, 41
Inorganic colorants, 89–93

International Association of Color Manufacturers, 30
Interpretation of color data by human brain, 8
Iridoid pigments, 86–88, 134–135
Iron oxide, synthetic, 28, 91, 102, 134–135
Isobetanidin, 62
Isobetanin, 63
Isocarthamin, 78
Isoprebetanin, 63

Joint FAO/WHO Expert Committee on Food Additives (JECFA), 26, 83, 85
Judd-Hunter system, *see* Measurement of color, L a b system

Kermes, 75, 134–135
Kermes ilicis, 75
Kermococcus vermilis, 75
Kubelka-Munk equations, 17

Labeling, of colorants, 29, 30
Lac, 75–76, 136–137
Laccifera lacca, 75, 95
Lakes, 30, 40–41
  in baked goods, 98, 99
  in confections, 119–120, 122
  in dry-mix beverages, 108
  manufacturing process, 40
  in pet foods, 102–103
Laminaria, species, 94
Lecithin, 99
Legal status of colorants, information sources, 31
Lutein, 43, 48, 129
Lycopene, 43, 44, 48–49, 50, 136–137
Lycopersicon esculentum, 48

Macrocystis species, 94
Malvidin, 56
Margaroides polonicus, 74
Market for colorants, 5
Martius Yellow, 3
Measurement, of color
  calculation, 10
  CIE XYZ system, 10, 11, 15, 17, 22
  color coordinates, 9–10
  color standards, 8
  correlations between systems, 18
  interpretation, 17–22
  L a b system, 11, 15, 17, 18
  L* a* b* system, 11, 15, 17, 18, 19
  L* C* H* system, 11, 15, 19, 20
  Munsell system, 8, 9, 15, 19
  sample presentation, 15–17
  by spectrophotometry, 9–12
  standard observer curves, 10
  visual systems, 8
Medicago sativa, 69

Metals, effect on colorants, 67, 68, 69, 74–75, 78, 127
Molds, colorants source, 50
Monascin, 79
Monascorubin, 80
Monascorubramine, 80
Monascus colorants, 136–137
  applications, 80–81
  chemistry, 79–80
  production, commercial, 80
Monascus species, 80
Mood, and colors, 4
Munsell system, of color measurement, 8, 9, 15–16, 19
  color solid, 15–16

Naphthoqinones, 72
Natural colorants, 5, 27, 129. *See also* Exempt colorants
  in baked goods and cereals, 99–102
  in beverages, 108–112
  concept not accepted by FDA, 29, 30
Natural Orange 4, *see* Annatto
Neobetanidin, 62
Neoxanthin, 43, 49
Nettles, source of chlorophylls, 69
Neurosporene, 44
No observed adverse effect level (NOAEL), 24
No observed effect level (NOEL), 24
Nomenclature, commission on (1980), 67
Nopalea coccinellifera, 74
Norbixin, 44, 109
Nutraceuticals, 49
Nutritional dietary supplements, colorants as, 89, 91
Nutritional Labeling and Education Act (1990), 29

Octopus ink, 95, 136–137
Optical signature, of a colorant, 84
Opuntia species, 63, 74
Orange B, 36, 136–137
Organic colorants, miscellaneous, 93–95

Packaging materials, 41, 89, 119
Palm oil, 49, 136–137
Panning equipment, 120
Paprika, 28, 46–47, 136–137
Paprika oleoresin, 28, 47, 99, 101, 118, 136–137
Patents, for food colorants, 32
Peonidin, 56
Perkin, W. H., and synthetic colorants (1856), 2, 33
Pet foods, 102–103, 104–105. *See also* Animal feeds
Petunidin, 56

pH, influence on colorants, 127
  anthocyanins, 58, 59, 60
  in beverages, 108, 109, 110, 111
  carminic acid, 75
  in confections, 118
  in yogurt, 113
Phaffia rhodozyma, 50, 129
Phenolic acids, in grapes, 56
Pheophorbide, 67, 69
Pheophytin, 67, 69
Phloxene B, *see* FD&C Red No. 28
Phycobilins, 70–72, 136–137
Phycocyanins, 70, 71, 72, 101, 129
Phycocyanobilin, 71, 72
Phycoerythrins, 70, 72
Phycoerythrobilin, 71
Phycomyces blakesleeanus, 50
Phytofluene, 44
Phytolacca americana, 62, 63
Phytolaccatoxin, 62
Pigment content, measurement, 22
Pleuroncodes planiples, 50
Pokeberry, 62, 63
Pokeweed, 61–62
Polishing candy tablets, 121
Polyphenols, brown, *see* Brown polyphenols
Porphyra species, 94
Porphyridium cruentum, 72
Porphyrophera species, 74
Poultry feed, colorants, for, 47, 48, 49
Prebetanin, 63
Precarthamin, 78, 79
Preservatives, 126
Primary colors, 20
Protein, in pigment, 72
Provisional listing of colorants and lakes, 41
Purity
  of color, 10, 30
  of color additives, 85
Purpurinidin fructoglucoside, 59

Quality, factors affecting, 126–128
Quercus coccifera, 75
Quinoline Yellow (in Europe), 36
Quinoline Yellow (in the United States), *see* D&C Yellow No. 10

Red cabbage, 55, 56, 59, 110, 132–133
Red sweet potato, 59
Reference toxicity dose (rTD), 25, 26
Regulation of colorants
  in the European Union, 26–27
  food inspection decisions (FIDs), 2, 3, 33
  history, in the United States, 2, 3, 23, 33, 34
  summary of U.S. regulations, 30

Rhizopus species, 88
Rhodomenia palmata, 94
Riboflavin, 28, 94, 136–137
Rods, in human eye, 7
rTD, *see* Reference toxicity dose
Rubropunctamine, 80
Rubropunctatin, 80

Safety, of colorants
  and regulations, 23–26, 34, 36
  testing for, 23–25, 85
Safflor red, see Carthamin
Safflor yellow, A and B, 78, 79
Safflowers, 78
Saffron, 28, 46, 136–137
Saponified marigold extract, 48
Scientific Committee for Food (European Union), 26
Scleranthus perennis, 74
Shellac, 95
Silicon dioxide, 92–93, 136–137
Silver, 92, 136–137
Society of Dyers and Colorists, 27
Solubility, of colorants, 38, 107
Spectrophotometry, 9–11
Spectrum, visible, 9, 10, 21
Spirulina species, 72, 129
Spongiococcum species, 94
Squid ink, 95, 136–137
Stability of colorants, in foods, 110, 111, 118, 128
  anthocyanins, 58, 59
  FD&C colorants, 36, 38
  in storage, 110, 111, 128
Standards, for color measurement, 8
Sugars, 36, 56, 83, 99, 120
Sulfur dioxide, effect on colorants, 57, 59, 127
Sunset Yellow, *see* FD&C Yellow No. 6
Sweetness, perception influenced by color, 4
Synthetic colorants, 2, 33, 91, 94. *See also* Certified colorants *and* FD&C colorants
  anthocyanins, 59
  carotenoids, 51–52
  in the European Union, 27

Tagetes, 28, 30, 47–48, 136–137
  meal, 48
Tagetes erecta, 47
Talc, 91, 136–137
Tartrazine, *see* FD&C Yellow No. 5
Tea, as a colorant, 86, 136–137
Thea sinensis, 86
Theaflavin, 86
Thearubin, 86
Theobroma cacao, 86
Theta value, 17, 18, 20

Tinctorial power (strength), 84
  of anthocyanin extracts, 57
  of beet colorants, 64, 118
  of safrron, 46
  of turmeric, 78
Titanic Earth, 89
Titanium dioxide, 28, 89–90, 102, 103, 136–137
Tocopherol, 69
Toxicity, testing
  principles, 23–25
  studies, 25–26
Troubleshooting guides, 103–105, 114–116, 123
Turmeric, 28, 77–78, 99, 101, 103, 119, 136–137
Turmeric oleoresin, 28, 77, 78, 99, 100, 103, 136–137

Ultramarine Blue, 28, 90–91, 136–137
Ultramarine colorants, 90–91, 136–137
Urtica dioica, 69
U.S. Code of Federal Regulations, 30, 31–32, 40
U.S. Department of Agriculture (USDA)
  color measurement system, 8
  early regulation of colorants, 2
U.S. Environmental Protection Agency (EPA), 25, 31
U.S. Food and Drug Administration (FDA), 25
  and concept of natural colorants, 29
  listing of certified colorants, 33–34, 41
U.S. Food Inspection Decision 77 (1907), 27

Value (Munsell), 15, 19
Vegetable extracts, 93
Vegetable juice, 28, 136–137
Vegetable juice color additive, 64, 110–111
Violaxanthin, 43, 49, 50
Virtually safe dose, 25
Vitamins, source of colorants, 52, 69, 94, 112
Vulgaxanthin, 62

Water-soluble colorants
  in beverages, 107
  in confections, 117, 119, 121
  handling and preparing, 125–126
  in pet foods, 102
Wine, use of anthocyanins in, 59
Wrappers, food, colorants for, 40, 91, 95

Xanthophyll pastes, 49, 136–137

Zeaxanthin, 46, 48, 51, 129
Zinc oxide, 91, 136–137